Mechanical Properties of Polycarbonate

Series Editor
Piotr Breitkopf

Mechanical Properties of Polycarbonate

Experiment and Modeling for Aeronautical and Aerospace Applications

Weihong Zhang
Yingjie Xu

ISTE PRESS

ELSEVIER

First published 2019 in Great Britain and the United States by ISTE Press Ltd and Elsevier Ltd

ISTE Press Ltd
27-37 St George's Road
London SW19 4EU
UK

www.iste.co.uk

Elsevier Ltd
The Boulevard, Langford Lane
Kidlington, Oxford, OX5 1GB
UK

www.elsevier.com

British Library Cataloguing-in-Publication Data
A CIP record for this book is available from the British Library
Library of Congress Cataloging in Publication Data
A catalog record for this book is available from the Library of Congress
ISBN 978-1-78548-313-4

Contents

Preface

Polycarbonate is a kind of thermoplastic polymer with high transparency, high ductility, impact resistance and lightweightness. It has been widely used in transparent products of aeronautical and aerospace systems, including aircraft windshields, canopies and astronaut helmet visors. In order to design these products to satisfy complex service conditions, we should have a good understanding of the mechanical properties of polycarbonate that depend on a variety of factors such as strain rate, temperature and even the processing conditions. From the viewpoint of experiments, it is almost impossible to have a comprehensive understanding of the mechanical behavior of polycarbonate products and their evolution rules under various loading conditions. A promising way is to develop numerical modeling techniques.

With the support of the National Key Research Program and the National Natural Science Foundation of China, our research group has been devoted to the development of numerical modeling techniques for studying the mechanical properties of polycarbonate with emphasis on aeronautical and aerospace applications. The main content of this book covers experimental characterization, material modeling, finite element simulation as well as the integrated analysis and design method for polycarbonate. This book contains six chapters:

– the Introduction briefly presents the properties, processing and applications of polycarbonate. The purpose and layout of this book is also presented;

– Chapter 1 presents the experiment facilities and methods used to characterize the mechanical properties of polycarbonate;

– Chapter 2 focuses on the constitutive modeling of polycarbonate and the implementation of the constitutive model in finite element tools;

– Chapter 3 focuses on the finite element simulation of polycarbonate products under impact;

– Chapter 4 presents an integrated simulation framework that incorporates the inhomogeneous yield stress distribution induced by injection molding into the mechanical simulations of polycarbonate products;

– Chapter 5 presents a process optimization methodology to improve the mechanical performance of polycarbonate products under impact loading.

This book is not indented to capture all the significant contributions that have been previously reported in the literature. It can be considered as a complementary work of existing books about polymer mechanics and processing. This book can be used as a guideline for graduate students and research engineers who wish to learn the basic experimental and modeling techniques for characterizing and designing polycarbonate products. Readers need the technical background of mechanics and finite element method.

Weihong ZHANG
Yingjie XU
April 2019

Introduction

I.1. Mechanical properties of polycarbonate

Polycarbonate is a typical thermoplastic polymer whose molecular chains are associated with intermolecular attractive forces (Ward and Hadley 1993). These forces weaken rapidly when the temperature increases, which causes the solid polycarbonate to turn into a viscous liquid. Thus, polycarbonate can be reshaped by heating it and it is typically used to produce products through various processing techniques such as injection molding, compression molding, calendering and extrusion (Gedde 2013).

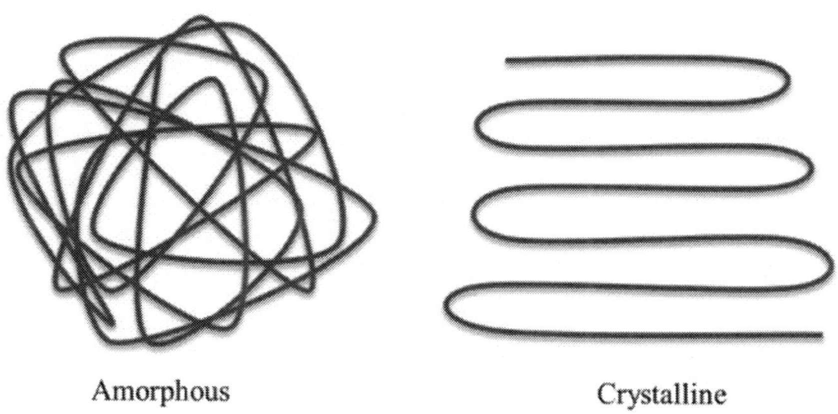

Figure I.1. *Molecular chains in amorphous and crystalline polymers*

The thermoplastic polymer can be further divided into two types: amorphous polymer and crystalline polymer. The main difference between amorphous and crystalline polymers lies in their molecular arrangement. The former has a molecular structure that has no organized pattern, whereas the latter has an ordered molecular structure, as shown in Figure I.1. The atactic molecular structure causes weak intermolecular forces in the amorphous polymer. Therefore, the amorphous polymer has low density and chemical resistance compared to the crystalline polymer. In addition, crystalline structures reflect the light strongly because of their large crystallites, whereas the absence of crystallinity in the amorphous polymer generates high optical clarity.

Polycarbonate has been widely used in engineering applications over the last decades. This growing application is partly derived from the ease with which it can be formed into any shape, and is partly because of its generally excellent performance. As one of the most commonly used amorphous polymers in engineering, polycarbonate was first developed in 1953 at Bayer AG in Germany and at General Electric in the USA independently.

As shown in Figure I.2, a polycarbonate molecule is composed of a bisphenol A group and a carbonate group. Bisphenol A contains two aromatic rings and constitutes the stiff backbone of polycarbonate. It also contributes to the inability of crystallization and thus gives particular transparency to polycarbonate.

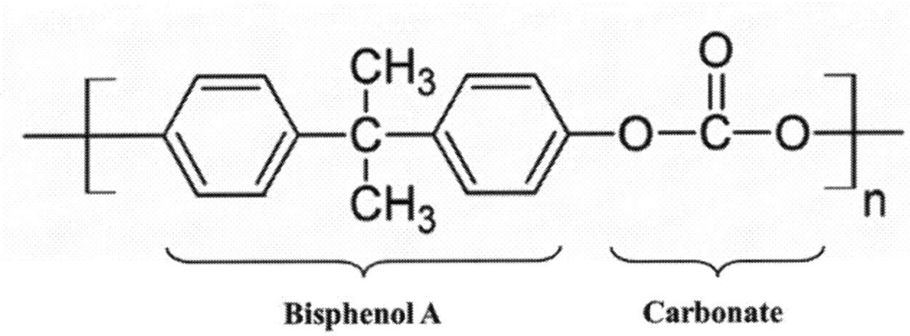

Figure I.2. *Molecular structure of polycarbonate*

The main physical properties of polycarbonate are listed in Table I.1[1]. These properties often increase with the increase in temperature. Clearly, polycarbonate is highly transparent to visible light with good light transmission (up to 93% of visible light) compared to many types of glasses. It also exhibits low moisture and water absorbing capacity so that products made of polycarbonate have a good dimensional stability.

Density (g/cm^3)	Light transmission (%)	Water absorption over 24 hours (%)	Moisture absorption at equilibrium (%)
1.20–1.22	80–93	0.1%	0.16–0.35%

Table I.1. *Physical properties of polycarbonate*

Table I.2 shows the mechanical properties of polycarbonate at room temperature[1]. Compared with a majority of engineering materials, polycarbonate has a large elongation at break (80–150%). It can thus undergo large plastic deformation without cracking or breaking. Moreover, polycarbonate has an outstanding impact resistance. As shown in Figure I.3[2], the notched Izod impact strength of polycarbonate is about 40 times larger than that of acrylic and 10 times larger than that of PETG, which are two commonly used transparent thermoplastics

Young's modulus (GPa)	Tensile strength (MPa)	Elongation at break (%)	Poisson's ratio	Notched Izod impact strength (J/cm)
2.0–2.4	55–75	80–150	0.37	22.5–30.0

Table I.2. *Mechanical properties of polycarbonate*

1 https://en.wikipedia.org/wiki/Polycarbonate.
2 https://www.curbellplastics.com.

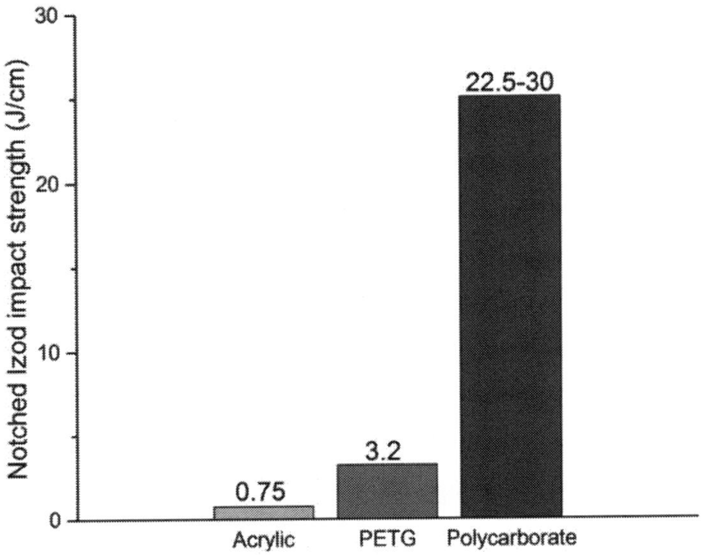

Figure I.3. *Impact resistance of polycarbonate compared with acrylic and PETG*

The thermal properties of polycarbonate are listed in Table I.3[3]. Polycarbonate has a glass transition temperature of about 145°C. This means that it softens gradually above this temperature and can maintain good mechanical properties below its glass transition temperature. The melting temperature of polycarbonate is about 265°C, at which it changes from the solid state to the liquid state. Its resistance to temperature changes is relatively high among amorphous polymers.

Glass transition temperature (°C)	Melting temperature (°C)	Working temperature (°C)	Thermal conductivity (W/(m·K))	Thermal expansion coefficient (10^{-6}/K)
145	265	−70–130	0.19–0.22	65–70

Table I.3. *Thermal properties of polycarbonate*

3 https://en.wikipedia.org/wiki/Polycarbonate.

I.2. Processing of polycarbonate

Polycarbonate is usually obtained in granular form as a raw material, as shown in Figure I.4. According to the shape of the final product, polycarbonate granules are first heated to a temperature higher than the melting point. The melted polycarbonate is then extruded or pressured into a mold to give the desired shape (Potsch 1995).

Figure I.4. *Polycarbonate granules*

– **Extrusion:** melted polycarbonate is passed through a mold to produce the final shape. Then, the melt is cooled rapidly. Long pipes and sheets are usually produced by this process.

– **Injection molding:** melted polycarbonate is pressed into a mold with the predefined shape of the final product. The melt is then cooled inside the mold. This process is ideal for products with complex geometry.

Due to the capability of manufacturing products with complex geometries, injection molding has been the most commonly used processing method for polycarbonate. Figure I.5 shows a schematic of a typical injection molding machine. Granular raw material is fed through a hopper into a heated barrel with a reciprocating screw that delivers the raw material forward, and mixes and homogenizes the thermal and viscous distributions of polycarbonate. When enough materials are gathered, they are forced at a high pressure and velocity into the mold cavity. Once the screw reaches the transfer position, the packing pressure is applied on the mold to complete mold filling and to compensate for thermal shrinkage. The injection pressure is applied at the cavity entrance until the solidification is complete. The material within the mold is cooled by temperature-controlled water or oil circulating in cooling lines. Once the required mold temperature or the holding time is achieved, the mold is opened and the product is obtained.

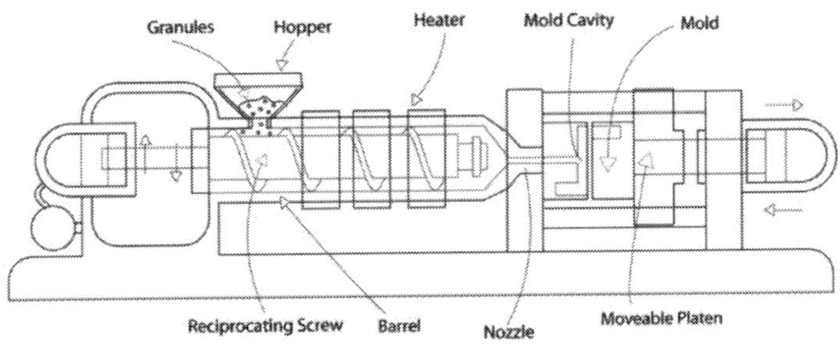

Figure I.5. *Schematic of a typical injection molding machine*

I.3. Engineering applications of polycarbonate

Due to its inherent features such as lightweightness, impact resistance, high transparency and heat resistance, polycarbonate has been used in a wide variety of fields[4]:

– **Electronic components:** the heat-resistant and electrical-insulating properties make polycarbonate a perfect choice in the electrical and electronic industry. The most common applications include switching relays, sensor parts, LCD sections, dielectric in high-stability capacitors, cell phone and lamp covers. Closely related to this is the optical industry where polycarbonate is used to make optical discs as a storage medium. These include CDs, CD-ROMs and DVDs.

– **Civil constructions:** considering its properties of transparency, impact resistance, lightweightness and thermal insulation, polycarbonate is an ideal choice for civil constructions. Typical applications include the safety and security glazing of buildings, overhead glazing and greenhouse covering.

– **Automotive:** polycarbonate is the dominant material for making transparent parts in the automotive industry. Headlamp lenses and windows of automobiles and trains are commonly made of polycarbonate. Windscreens in some small motorized vehicles such as motorcycles and ATVs are also made of polycarbonate.

4 https://en.wikipedia.org/wiki/Polycarbonate.

– **Protection components:** polycarbonate is widely used in projectile protection viewing and lighting applications owing to its high impact resistance. It is used to make many types of safety glasses and goggles, helmet visors, riot shields, etc. It is also used in bullet-resistant windows in automobiles and barriers in banks.

– **Aeronautical and aerospace applications:** polycarbonate is commonly used in the manufacture of military and civilian aircraft windshields, canopies and observation windows. Figure I.6 shows the typical applications, including the canopy of the F-16 fighter, windshields of the Bell 206 series helicopter and the Pipistrel electric aircraft. Polycarbonate is widely used in aerospace applications to make the optical visors of the astronaut's helmet. Figure I.7 shows the applications of polycarbonate in optical visors of helmets for astronauts of NASA's Gemini, Apollo in the USA and CASC's ShenZhou in China.

(a) Canopy of F-16 (b) Windshield of Bell 206B helicopter (c) Windshield of Pipistrel electric aircraft

Figure I.6. *Polycarbonate canopies and windshields in aircrafts*

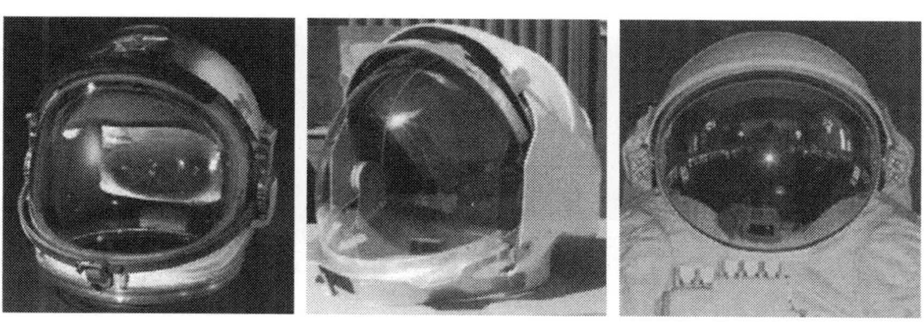

(a) Gemini (b) Apollo (c) ShenZhou

Figure I.7. *Polycarbonate visors of astronaut helmet*

I.4. Challenges in aeronautical and aerospace applications

Polycarbonate is considered a type of commonly used material for transparent products in aeronautical and aerospace engineering. Compared with electronic, civil construction, automotive and other civilian applications, the service conditions become more complex and challenging.

First, the threats are unavoidable in terms of high velocity impacts. For instance, collision between birds and aircrafts, known as a bird strike, has been one of the most important safety threats faced in the aeronautical industry (Abrate 2016). Among the different parts of an aircraft, the exposed windshield and canopy are more prone to bird strikes. The impact velocity of bird strikes could reach 250–350 km/h, depending on the take-off and landing speed of the aircraft (bird strikes usually occur during the take-off and landing stages). The strain rate of such an impact is usually more than 10^3. Bird strikes can result in the surface cracking or perforation of the windshield and canopy, as shown in Figure I.8. Therefore, polycarbonate windshield and canopy must be capable of withstanding high-speed bird strikes.

(a) Damage in F-16 aircraft canopy

(b) Damage in a Cessna aircraft windshield

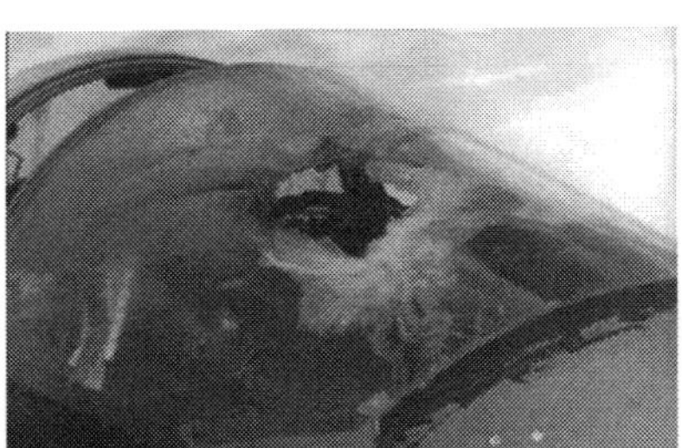

(c) Perforation in a Harrier Jet aircraft canopy

(d) Damage in Black Hawk helicopter windshield

Figure I.8. *Bird strike-induced damages on aircraft windshield and canopy. For a color version of this figure, see www.iste.co.uk/zhang/polycarbonate.zip*

Similarly, the visor of an astronaut helmet has to withstand the impact of small-sized debris to ensure the safety of astronauts. Protection against impacts from space debris which cannot be eye-tracked is now a routine in the design of aerospace products. For example, the impact energy of the debris that is 1 mm in size could reach 71 J[5]. This value sharply rises to 1910 J when the size of the debris increases to 3 mm.

Second, a large range of service temperatures should be taken into account. An aircraft can experience temperatures ranging from ground temperature to nearly -60°C as it flies from the ground to an altitude of 12,000 m. The temperature difference can reach more than 100°C and is even larger in aerospace applications. For example, the temperature of dayside of the international space station is about 120°C, while the temperature of nightside can be as low as -157°C. The temperature inside the space shuttle maintains about 20°C. Therefore, astronauts will experience a large temperature difference when they do extravehicular activities.

In order to design the polycarbonate products against catastrophic impact failure, we should have a clear understanding of the mechanical behavior of polycarbonate at large strain rates and temperature ranges to accurately predict how the failure occurs. This is quite challenging since the mechanical behavior of a polycarbonate product depends on a variety of factors. The details are discussed below.

Dependence on strain rate and temperature

Mechanical properties of polycarbonate present a strong dependence on strain rate and temperature. Figure I.9 presents the stress–strain responses of polycarbonate (Lexan 141R, provided by GE) under uniaxial compression loading at various strain rates and temperatures (Wang *et al.* 2016). As shown, both yield stress and post-yield behaviors are notably sensitive to the strain rate and temperature. However, mechanical characterizations of polycarbonate products in a purely experimental setting are expensive and time-consuming.

5 http://www.aerospace.org/cords/all-about-debris-and-reentry/debris-impacts-in-orbit.

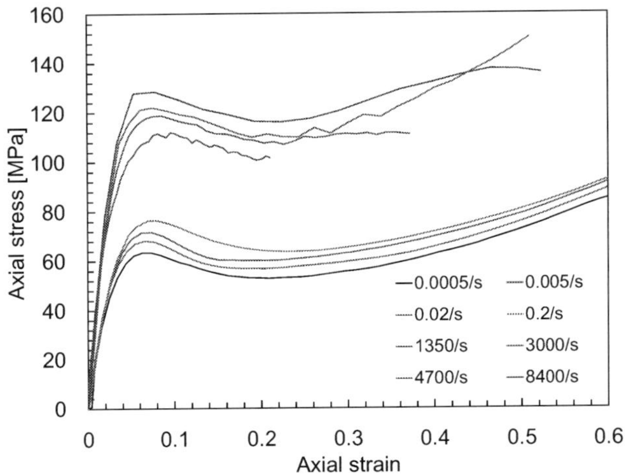

(a) Strain rate dependence

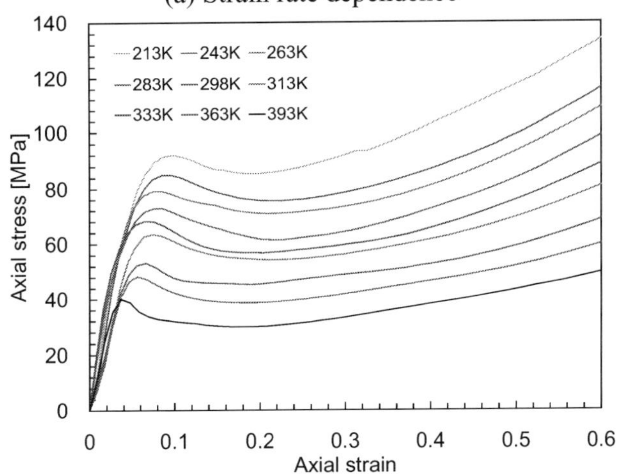

(b) Temperature dependence

Figure I.9. *Strain rate and temperature-dependent responses of polycarbonate. For a color version of this figure, see www.iste.co.uk/zhang/polycarbonate.zip*

Processing-affected yield stress

It is well-known that the mechanical properties of a polymer are influenced by its processing conditions. A typical example is the crystalline polymer, where the flow and temperature history experienced during processing results in an inhomogeneous, anisotropic crystalline morphology. As a result, the injection-molded products of crystalline polymers display different mechanical behaviors dependent on position and orientation (e.g. tough parallel to flow and brittle in the perpendicular direction). For polycarbonate and other amorphous polymers, the effects of flow-induced orientation during injection molding are much less pronounced. However, the thermal history experienced upon solidification from the melt has a remarkable influence on the mechanical properties of polycarbonate (Govaert *et al.* 2005). At the cooling stage of an injection molding process, the uncoiled molecules are pulled by intermolecular forces to return to an equilibrium state with random orientation, which is called relaxation. Obviously, variations in the cooling rate would result in different relaxations and consequently affect the mechanical properties of the molded product. The studies of Govaert *et al.* (2005), Engels *et al.* (2008 and 2009), Xu *et al.* (2015 and 2016) have proved that the processing thermal history has an influence on the yield stress of molded polycarbonate products.

In a practical injection molding process, the thermal history usually depends on processing parameters such as mold temperature and cooling time. Figure I.10 shows the experimental results Xu *et al.* (2015) of the yield stress of polycarbonate specimens processed at different mold temperatures. It can be observed that the variations in yield stress result in different failure modes for polycarbonate (Dar *et al.* 2017), i.e. ductile fracture in low mold temperature (60°C) and brittle fracture in high mold temperature (80 and 120°C).

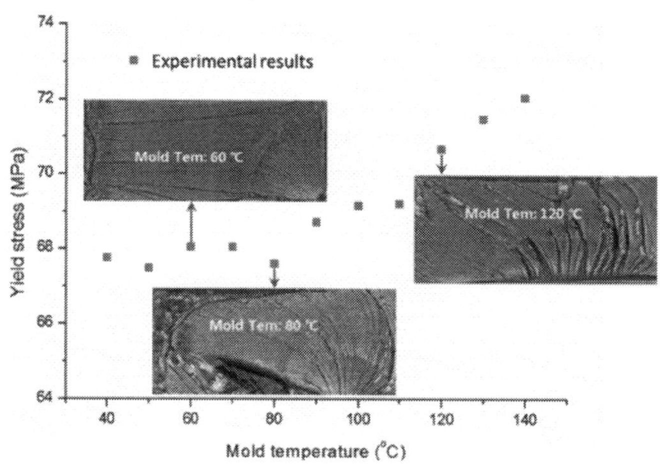

Figure I.10. *Yield stress of polycarbonate specimens processed by different mold temperatures*

In fact, even for a product with simple geometry, the cooling is non-uniform so that the inhomogeneity of yield stress produces throughout the product. For example, during the processing of a polycarbonate plate, cooling is fast near the plate surface in contact with the mold cavity but slow at the center. Therefore, the yield stress of the molded plate exhibits a gradient distribution along the thickness (Engels *et al.* 2009), which is consistent with the observation of the microstructural morphology of the plate shown in Figure I.11.

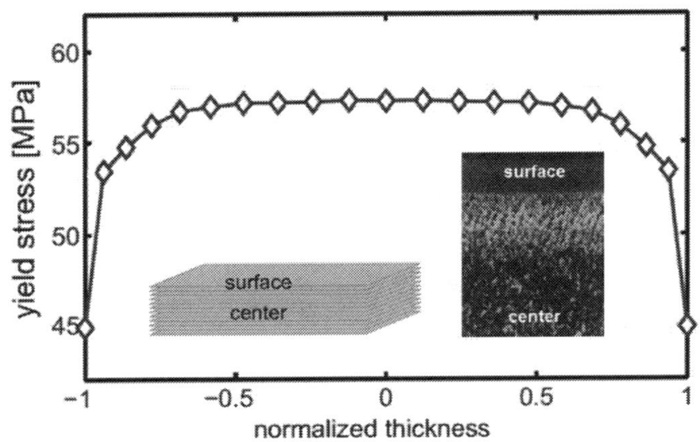

Figure I.11. *Gradient distribution of yield stress along the plate thickness*

Processing-induced warpage

Warpage is considered one of the most serious defects in injection-molded thin-walled products. It can be defined as a dimensional distortion in a molded product after its ejection from the mold. Warpage is mainly caused by the non-uniform shrinkage of material during processing[6]. Figure I.12 shows the basic schematic of polycarbonate shrinkage. In the equilibrium state, polycarbonate has a random and entangled molecular orientation. When it is melted, intermolecular forces are weakened and the entangled molecules are uncoiled. At the cooling stage of processing, uncoiled molecules are pulled by intermolecular forces to return to the equilibrium state with random orientation. However, non-uniform shrinkage induces

6 https://www.autodesk.com/industry/manufacturing/resources/injection-molding/causes-of-warpage.

internal stresses and thus warps the product. Non-uniform shrinkage often appears in the injection molding process.

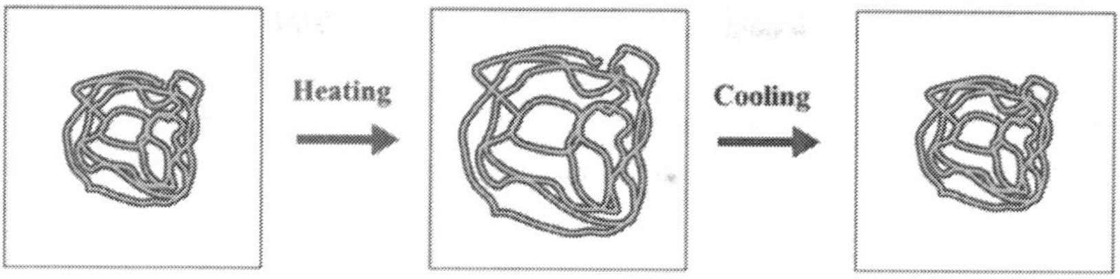

Figure I.12. *Schematic of the shrinkage of amorphous polymers*

Due to the warpage, the molded product would be forced to deform from the "warped shape" to the "designed shape" in the assembling process, as shown in Figure I.13. As a result, residual stresses are thus generated within the assembled product and influence the mechanical response and functionality of the product.

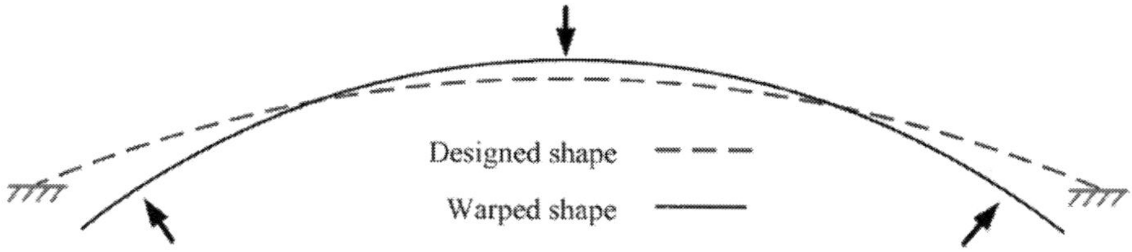

Figure I.13. *Schematic of the forced deformation in the assembling of the warped product*

I.5. Purpose and layout of this book

The discussions presented above indicate that the mechanical behavior of a polycarbonate product depends on a variety of factors such as strain rate, temperature and even the processing conditions. Based on a purely experimental setting, it is almost impossible to have a comprehensive understanding of the mechanical behavior of these products and their evolution rules under various loading conditions, temperatures and processing parameters. A promising way is to develop numerical modeling techniques.

In general, two types of numerical modeling approaches are suitable to study the mechanical behavior of polymers, i.e. microscopic and macroscopic approaches. The microscopic approach focuses on the molecular structure whose evolution is related to the macroscopic behavior. Although the microscopic approach helps to gain a better understanding of the underlying mechanism of the material, it is rather cumbersome and time-consuming in engineering applications. Comparatively, the macroscopic approach is an appropriate choice in practice. Polymers are modeled within the framework of continuum mechanics based on the experimental observation. The finite element (FE) method is further used to simulate the mechanical behavior.

At present, several commercial FE tools (e.g. Moldflow and Moldex3D for processing the simulation of polymer products, and Abaqus and LS-DYNA for mechanical simulation) are available. However, the existing FE tools rarely contain material models suitable for characterizing the mechanical properties of polycarbonate within the large strain rates and temperature ranges. Meanwhile, the influence of processing simulation on mechanical simulation is not taken into account. In other words, processing simulation is only limited to the evaluation of shrinkage, warpage, etc. that are not considered in the evaluation of mechanical behavior of the molded product.

Therefore, the following questions should be addressed when designing and manufacturing a polycarbonate product:

– What about experiments needed for characterizing the mechanical properties of polycarbonate? How to conduct these experiments?

– How to model the strain rate and temperature-dependent behavior of polycarbonate? How to implement the developed model in FE tools?

– What about the influence of processing conditions on the mechanical behavior of a molded product? How to incorporate the processing influence into a mechanical simulation? How to design processing parameters to improve the mechanical performance of a product?

This book attempts to develop an integrated methodology that combines the processing and mechanical simulations for polycarbonate products, as shown in Figure I.14. It can be considered as a complementary work of existing books about polymer mechanics (Ward and Hadley 1993;

Gedde 2013; Bergstrom 2015) and processing (Barone *et al.* 1989; Potsch 1995; Zhou 2012; Tadmor and Gogos 2013). To this end, the topics in this book cover experimental characterization, material modeling, FE simulation as well as the integrated analysis and design method for polycarbonate.

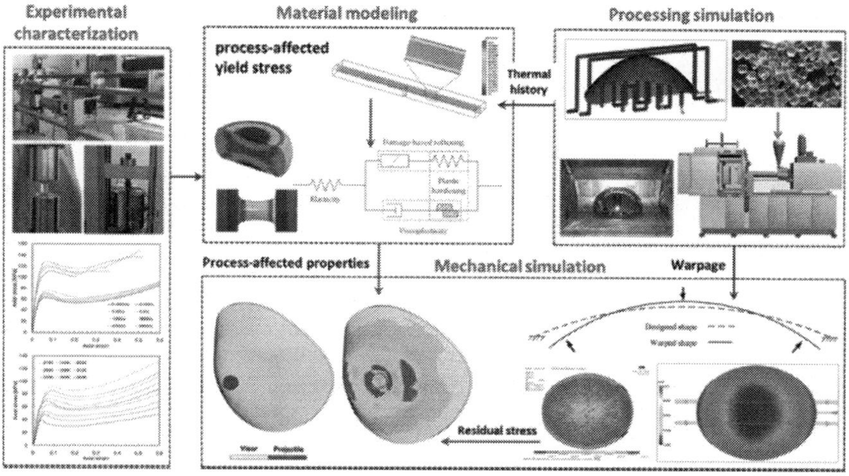

Figure I.14. *Schematic of the integrated methodology. For a color version of this figure, see www.iste.co.uk/zhang/polycarbonate.zip*

Chapter 1 presents the experiment facilities and methods used in characterizing the mechanical properties of polycarbonate. More specifically, it presents in detail the discussions about the realization of the strain rate- and temperature-dependent experiments.

Chapter 2 focuses on the constitutive modeling of polycarbonate. Some of the most commonly used models, including simple linear viscoelastic models and advanced nonlinear viscoplastic models, are presented. Examples are provided to show how to implement the constitutive models in FE tools.

Chapter 3 focuses on the FE simulation of a polycarbonate product under impact. Typical aeronautical and aerospace products such as aircraft windshields and astronaut helmet visors are presented.

Chapter 4 focuses on the development of an integrated computing framework that incorporates processing-affected yield stress into the mechanical simulation of polycarbonate products. It explains in detail how to predict yield stress after processing and how to combine the prediction results with constitutive modeling and FE simulation.

Chapter 5 focuses on the optimization method of processing parameters to improve the mechanical performance of polycarbonate products. The integrated computing framework presented in Chapter 4 is further extended to take into account warpage-induced residual stress. An approximation model is developed and combined with a particle swarm algorithm to realize the optimization procedure.

I.6. References

Abrate, S. (2016). Soft impacts on aerospace structures. *Progress in Aerospace Sciences*, 81, 1–17.

Barone, M.R., Castro, J.M. and Ellson, R.N. (1989). *Fundamentals of Computer Modeling for Polymer Processing*. Hanser, Munich.

Bergstrom, J.S. (2015). *Mechanics of Solid Polymers: Theory and Computational Modeling*. William Andrew, Amsterdam.

Dar, U.A., Xu, Y.J., Zakir, S.M. and Saeed M.-U. (2017). The effect of injection molding process parameters on mechanical and fracture behavior of polycarbonate polymer. *Journal of Applied Polymer Science*, 134(7).

Engels, T.A.P., Govaert, L.E. and Meijer, H.E.H. (2008). Quantitative prediction of mechanical performance of polymer products directly from processing conditions. *American Institute of Physics Conference Proceedings*, 1027(1336).

Engels, T.A.P., van Breemen, L.C.A., Govaert, L.E. and Meijer, H.E.H. (2009). Predicting the long-term mechanical performance of polycarbonate from thermal history during injection molding. *Macromolecular Materials and Engineering*, 294(12), 829–838.

Gedde, U.W. (2013). *Polymer Physics*. Springer Science & Business Media, Berlin.

Govaert, L.E., Engels T.A.P., Klompen, E.T.J., Peters, G.W.M. and Meijer, H.E.H. (2005). Processing-induced properties in glassy polymers. *International Polymer Processing*, 20(2), 170–177.

Potsch, G. (1995). *Injection Molding: An Introduction*. Hanser Publishers, Munich.

Tadmor, Z. and Gogos, C.G. (2013). *Principles of Polymer Processing*. John Wiley & Sons, New York.

Wang, J., Xu, Y.J., Zhang, W.H. and Moumni, Z. (2016). A new damage-based model for the non-linear behavior of polycarbonate polymers. *Material & Design*, 97, 519–531.

Ward, I.M. and Hadley, D.W. (1993). *An Introduction to the Mechanical Properties of Solid Polymers*. John Wiley & Sons, New York.

Xu, Y.J., Lu, H., Gao, T.L. and Zhang W.H. (2015). Predicting the low-velocity impacted behavior of polycarbonate: Influence of thermal history during injection molding. *International Journal of Impact Engineering*, 86, 265–273.

Xu, Y.J., Lu, H. and Zhang, W.H. (2016). Processing-induced inhomogeneity of yield stress in polycarbonate product and its influence on the impact behavior. *Polymers*, 8(3), 72.

Zhou, H.M. (2012). *Computer Modeling for Injection Molding: Simulation, Optimization, and Control*. John Wiley & Sons, New York.

1

Experimental Studies of Mechanical Properties of Polycarbonate

The mechanical properties of polycarbonate exhibit a strong dependence on strain rate and temperature. Thus, a comprehensive study of such dependence is of great importance in the engineering design of polycarbonate products. The most common method to characterize the mechanical properties of polycarbonate is based on uniaxial tension and compression tests of a specimen in order to determine the stress–strain relation. For isotropic materials, the uniaxial stress–strain relation can be further extended to three-dimensional constitutive laws.

This chapter focuses on the experimental investigations of a wide range of strain rates and temperatures. According to the availability of experiment facilities and the real service environment of aeronautical and aerospace applications, the strain rate and temperature are considered within a range of 10^{-4}–10^{4} s^{-1} and −60–120°C respectively. The polycarbonate used in the experiment is Lexan 141R, supplied by SABIC Innovative Plastics Co., Ltd (China). The mass density is 1.2×10^3 kg/m^3, and the glass transition temperature (T_g) is 148°C which is determined by the dynamic mechanical thermal analysis (DMTA).

First, the strain rate-dependent properties are investigated using the quasi-static and high-rate dynamic tests. Both the compression and tension tests are implemented. Then, the temperature-dependent properties are investigated. Only quasi-static compression tests are conducted as a guideline for tests under tension and dynamic loading conditions.

1.1. Uniaxial compression tests at various strain rates

1.1.1. *Experimental setup of quasi-static uniaxial compression tests*

The quasi-static test is used at strain rates below 10^{-1} s^{-1}. As recommended in ASTM D695 (2015), cylindrical polycarbonate specimens with dimensions of 12.7 mm in diameter and 25.4 mm in length are used. They are cut from the extrusion-processed polycarbonate cylinders, and both ends are machined into smooth surfaces to reduce frictional effects during testing. Before experimental testing, the specimens are stored in a drying cabinet at room temperature (25°C) to release residual stress caused by the manufacturing process.

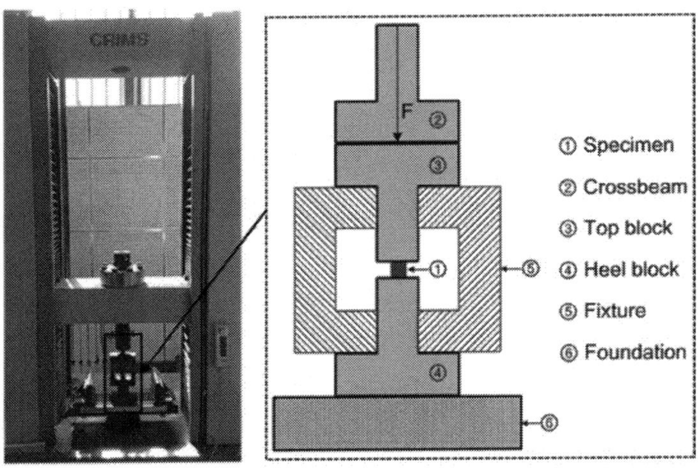

Figure 1.1. *Experimental setup for the quasi-static compression test*

Quasi-static uniaxial compression tests are performed on a CRIMS DNS 100 electromechanical universal testing system. A schematic of the experimental setup is shown in Figure 1.1. The specimen is sandwiched between two rigid blocks and compressed under the vertical load F at the desired strain rate.

During the compression test, the force–displacement data are first obtained and then converted into engineering stress–strain curves by computing engineering stress and engineering strain with respect to the undeformed reference length and diameter of the specimen. The engineering stress–strain curves are then transformed into true stress–strain curves to characterize the true mechanical properties of the tested material. It should be noted that the difference between the true and engineering stress–strain curves is not significant at low strains in the elastic regime. However, the difference is significant for large strains, as shown in Figure 1.2.

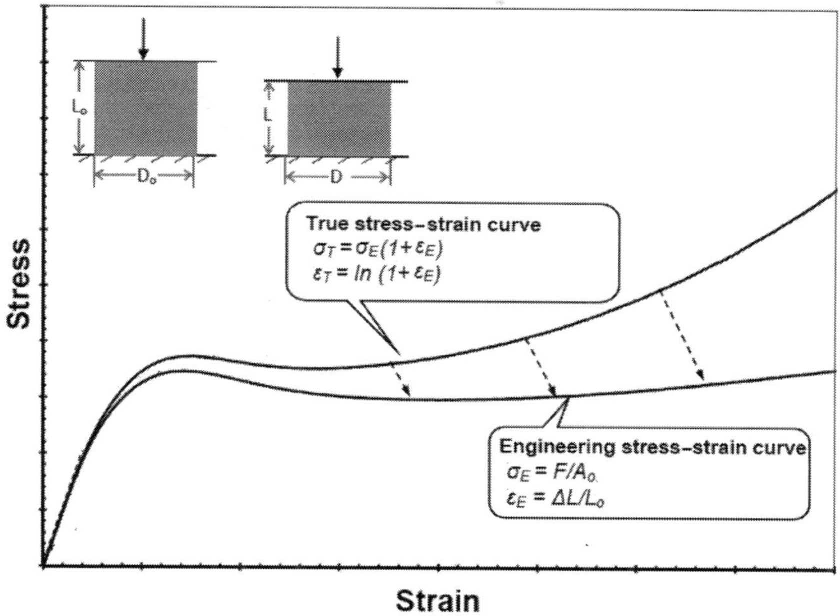

Figure 1.2. *Schematic curves of the engineering and true compressive stress–strain relations*

The engineering stress (σ_E) and engineering strain (ε_E) are defined as

$$\sigma_E = \frac{F}{A_o} \qquad [1.1]$$

$$\varepsilon_E = \frac{L - L_o}{L_o} \qquad [1.2]$$

where F is the applied load, L_o and A_o are the initial length and cross-sectional area respectively, and L is the instantaneous length of the specimen. The true stress (σ_T) is defined as the ratio of the applied load (F) to the instantaneous cross-sectional area (A) of the specimen:

$$\sigma_T = \frac{F}{A} \qquad [1.3]$$

Using the assumption of constant volume for plastic deformation and ignoring the elastic volume changes, we have

$$\text{AoLo} = \text{AL} \qquad [1.4]$$

True stress can thus be defined as

$$\sigma_T = \sigma_E (1 + \varepsilon_E) \qquad [1.5]$$

The relation between true strain (ε_T) and engineering strain (ε_E) is given by

$$\varepsilon_T = \int d\varepsilon = \ln\left(\frac{L}{L_o}\right) = \ln(1 + \varepsilon_E) \qquad [1.6]$$

1.1.2. *Experimental setup of dynamic uniaxial compression tests*

A dynamic test is often used to characterize the mechanical properties of materials at high strain rates, i.e. above 10^2 s^{-1}. The most common method used is the split Hopkinson pressure bar (SHPB) test system, also called the Kolsky bar (Chen and Song 2011). The SHPB test can be conducted under various loading conditions, including compression, tension, shear and torsion.

Although SHPB specimens are not standardized in dimension, cylindrical specimens of short length are often used to facilitate early dynamic equilibrium in a SHPB test. A large length is known to significantly attenuate the wave passing through the specimen (Chen *et al.* 2002). The ratio L/D (length L divided by diameter D) is usually set between 0.5 and 2 (Gray 1985a). In the present SHPB test, specimens with dimensions of 5.0 mm in diameter and 4.5 mm in length are used.

The SHPB test of polycarbonates is more complicated than that of metals because polycarbonates have low density and low modulus, ultimately resulting in low impedance, low amplitudes of transmitted pulses and an increase in the signal-to-noise ratio. To overcome these difficulties, different measures such as the pulse shaping technique and the use of low-impendence aluminum, titanium or polymeric bars can be used to perform the SHPB test (Gray 1985b; Wang *et al.* 1994).

Figure 1.3 shows the schematic of the SHPB setup that typically consists of a striker bar, an incident bar, a transmission bar and a strain acquisition system. All the bars are made of martensitic steel and have a diameter of 12.7 mm. The incident bar and the transmission bar are 1,200 and 1,000 mm in length respectively. The striker bar has a length ranging from 150 to 400 mm to produce different extents of strain range. An annealed copper pallet of diameter 5 mm is placed on the end face of the incident bar for pulse shaping. Figure 1.4 shows a real SHPB setup.

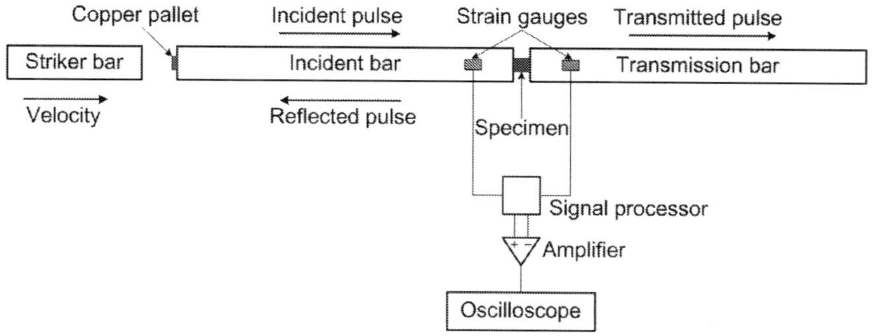

Figure 1.3. *Schematic of an SHPB setup*

During the test, frictional effects at the specimen–bar interfaces should be avoided as much as possible because they cause radial stress that in turn produces non-uniform deformation in the specimen. With this aim, petroleum jelly is often used to lubricate the specimen–bar interfaces in order to reduce interface friction.

Figure 1.4. *Real SHPB setup*

The striker bar is launched from a gas gun, which affects the incident bar and produces an elastic compressive stress wave traveling along the incident bar. The intensity of the stress wave is measured using a strain gauge attached to the incident bar. A portion of the stress wave propagates into the specimen when reaching the interface between the specimen and the incident bar, while the rest is reflected into the incident bar as tensile stress wave. The intensity of the reflected stress wave is measured again using the same strain gauge. The stress wave travels along the specimen and creates a compressive deformation within the specimen. Following this, the stress wave propagates into the transmission bar. The transmitted stress wave is then measured using a strain gauge attached to the transmission bar.

As the incident bar and the transmission bar are slender, one-dimensional wave propagation theory can be used to analyze the stress wave (Kolsky 1963). The wave propagation speed C_b in the bar can be calculated as (Gray 1985b)

$$C_b = \sqrt{E_b/\rho_b} \qquad [1.7]$$

where E_b and ρ_b are the elastic modulus and the density of the bar material respectively. Let us take martensitic steel bar as an example.

We have $E_b = 200$ GPa and $\rho_b = 7{,}780$ kg/m^3. The wave propagation speed is 5,070 m/s. In the present SHPB setup, the length of incident bar is $l_i = 1{,}200$ mm so that the stress wave will take 0.24 ms (l_i/C_b) to reach the specimen.

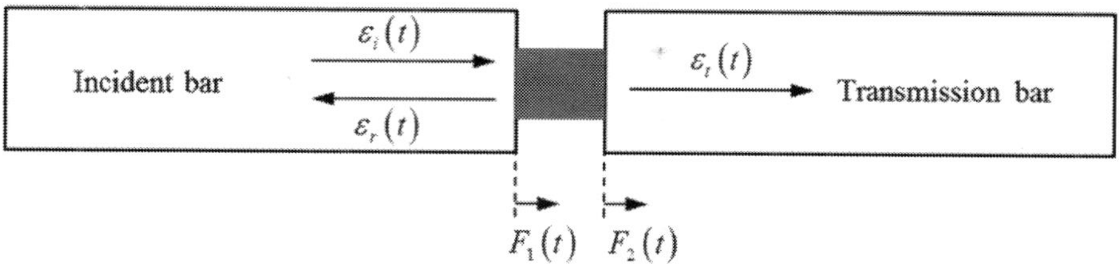

Figure 1.5. *Schematic of incident strain, reflected strain and transmitted strain*

Figure 1.5 shows the incident strain $\varepsilon_i(t)$, reflected strain $\varepsilon_r(t)$ and transmitted strain $\varepsilon_t(t)$ recorded as functions of time t using strain gauges located on the incident and transmission bars. The axial forces and displacements on the left end (contacted with the incident bar) and the right end (contacted with the transmission bar) of the specimen can be calculated from $\varepsilon_i(t)$, $\varepsilon_r(t)$ and $\varepsilon_t(t)$ using the following expressions (Gray 1985b):

$$F_1(t) = E_b A_b \left(\varepsilon_i(t) + \varepsilon_r(t)\right) \quad F_2(t) = E_b A_b \varepsilon_t(t) \qquad [1.8]$$

$$u_1(t) = C_b \int_0^t \left(\varepsilon_i(t) - \varepsilon_r(t)\right) dt \quad u_2(t) = C_b \int_0^t \varepsilon_t(t)\, dt \qquad [1.9]$$

where $F_1(t)$ and $F_2(t)$ are the forces, $u_1(t)$ and $u_2(t)$ are the displacements on the left end and the right end of the specimen respectively, and A_b is the cross-sectional area of the bar.

Furthermore, the engineering stress $\sigma_E(t)$, strain $\varepsilon_E(t)$ and strain rate $\dot{\varepsilon}_E(t)$ in the specimen can be obtained as

$$\sigma_E(t) = \frac{F_1(t) + F_2(t)}{2A_s} = \frac{E_b A_b}{2A_s}\left(\varepsilon_i(t) + \varepsilon_r(t) + \varepsilon_t(t)\right) \qquad [1.10]$$

$$\varepsilon_E(t)=\frac{u_1(t)-u_2(t)}{l_s}=\frac{C_b}{l_s}\int_0^t(\varepsilon_i(t)-\varepsilon_r(t)-\varepsilon_t(t))dt \qquad [1.11]$$

$$\dot{\varepsilon}_E(t)=\frac{C_b}{l_s}(\varepsilon_i(t)-\varepsilon_r(t)-\varepsilon_t(t)) \qquad [1.12]$$

where A_s and l_s are the cross-sectional area and the gauge length of the specimen respectively.

It is assumed that the equilibrium state exists in the SHPB test (Gray 1985b). This means that the forces on both ends of the specimen are equal with $F_1(t) = F_2(t)$, which gives $\varepsilon_i(t)+\varepsilon_r(t)=\varepsilon_t(t)$. In this sense, the measured values of $\varepsilon_i(t)$, $\varepsilon_r(t)$ and $\varepsilon_t(t)$ can be used to determine whether the specimen is in equilibrium or not during the experiment.

Based on the above equations, the engineering stress $\sigma_E(t)$, strain $\varepsilon_E(t)$ and strain rate $\dot{\varepsilon}_E(t)$ in the specimen can be obtained as

$$\sigma_E(t)=\frac{E_b A_b}{A_s}\varepsilon_t(t) \qquad [1.13]$$

$$\varepsilon_E(t)=-\frac{2C_b}{l_s}\int_0^t\varepsilon_r(t)dt \qquad [1.14]$$

$$\dot{\varepsilon}_E(t)=-\frac{2C_b}{l_s}\varepsilon_r(t) \qquad [1.15]$$

The true stress and true strain in the specimen can be easily calculated from the engineering stress and engineering strain using equations [1.5] and [1.6].

1.1.3. *Experimental results*

In the quasi-static compression test, polycarbonate specimens are monotonously compressed at strain rates ranging from 0.0005 to 0.2 s^{-1} at room temperature. The true stress–strain response of polycarbonate under uniaxial compression loading at a strain rate of 0.005 s^{-1} is shown in Figure 1.6.

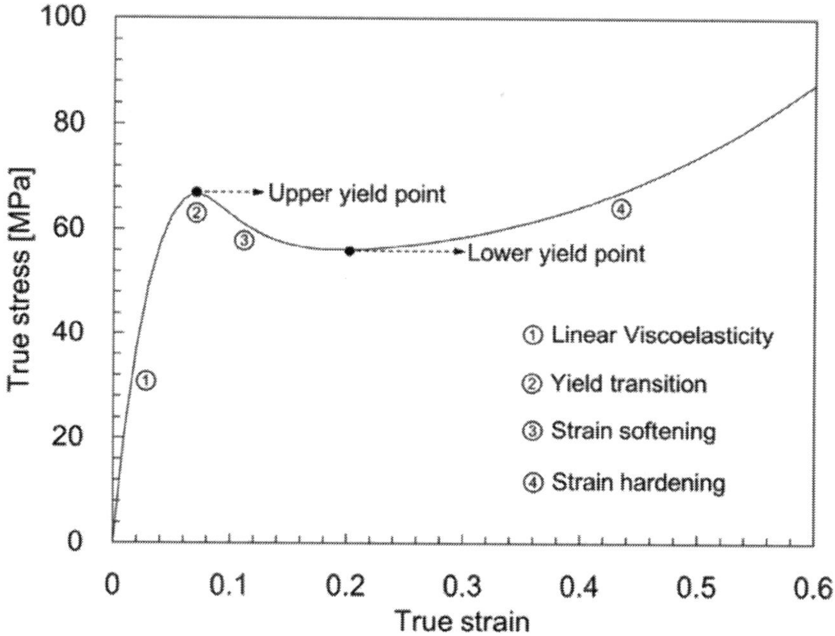

Figure 1.6. *True stress–strain response of polycarbonate under uniaxial compression loading at a strain rate of 0.005 s^{-1}*

During the quasi-static compression test, different deformation phases can be identified: initial linear elasticity, yield transition, strain softening and strain hardening. The linear elastic response takes place in the initial phase of loading. This response is principally known as the quasi-elastic response. With further loading, the nonlinear response becomes apparent until the attainment of the maximum stress, i.e. the upper yield point. Thereafter, a portion of specimen yields, flows without any further increase in stress and exhibits strain softening so that the flow stress reduces along with the further increase in the strain until a lower-stress plateau identified as the lower yield point is achieved. Therefore, the strain softening behavior can be represented by the difference between the upper yield stress and the lower yield stress.

Following this, the flow stress increases with increasing strain and eventually attains a level well beyond the initial yield stress. This region is commonly referred to as the strain hardening region.

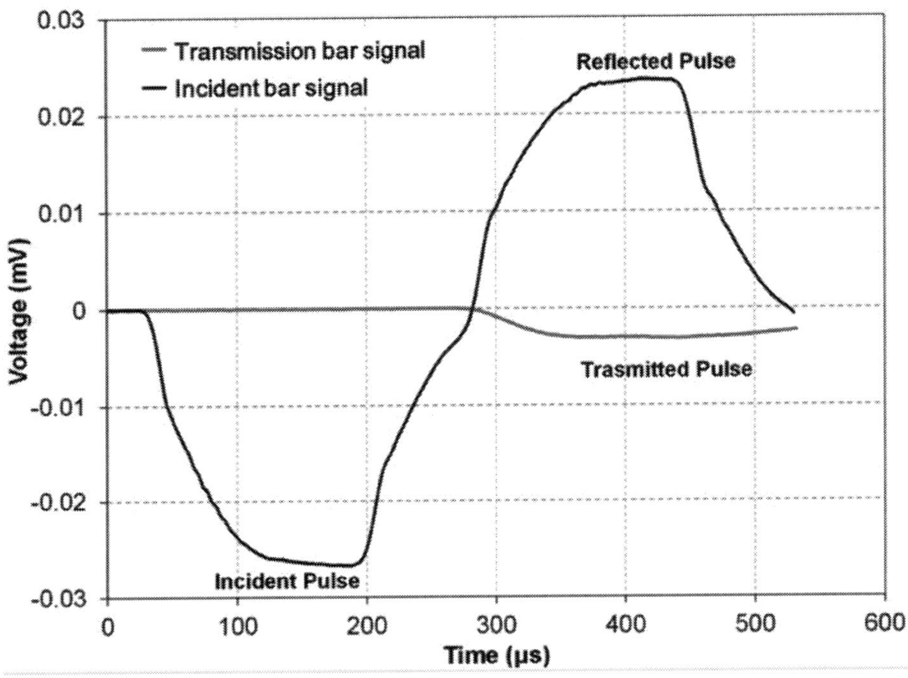

Figure 1.7. *Typical signals measured from the gauges on the bars. For a color version of this figure, see www.iste.co.uk/zhang/polycarbonate.zip*

The dynamic response of polycarbonate is determined at strain rates ranging from 1,350 to 8,400 s^{-1}. More specifically, at a strain rate of 1,350 s^{-1}, the strain signals of incident and transmission bars are extracted from the strain acquisition system, as shown in Figure 1.7. A nearly flat plateau at the reflected pulse indicates that the specimen is deformed at a constant strain rate.

The constant strain rate is one of the primary requirements in SHPB testing. The strain rate history is determined using equation [1.15], and the results of all the tests are shown in Figure 1.8. The flat portions show that almost a constant strain rate is achieved in each test.

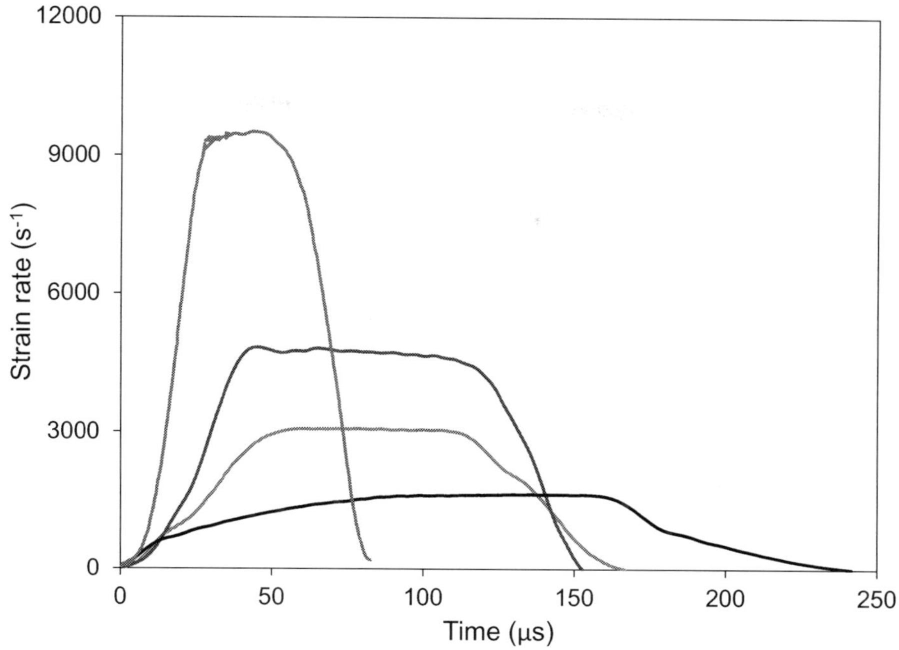

Figure 1.8. *Strain rate history of the specimen in each test. For a color version of this figure, see www.iste.co.uk/zhang/polycarbonate.zip*

Figure 1.9 shows the true stress–strain responses of polycarbonate over a wide range of strain rates (0.0005–0.2 s^{-1} in quasi-static testing and 1,350–8,400 s^{-1} in dynamic testing). The first observation is that the dynamic response of the material is distinct from the quasi-static one. From low to moderate strain rates, specimens deform in a ductile manner up to a strain of 0.60. In these tests, the stress–strain responses are similar and the stress–strain curves are relatively smooth. However, the yield stress and the plastic flow stress slightly increase with the increase in the strain rate. At high strain rates, the yield stress and the plastic flow stress increase dramatically. Moreover, fluctuations can be observed due to the effect of stress wave propagation on stress–strain curves.

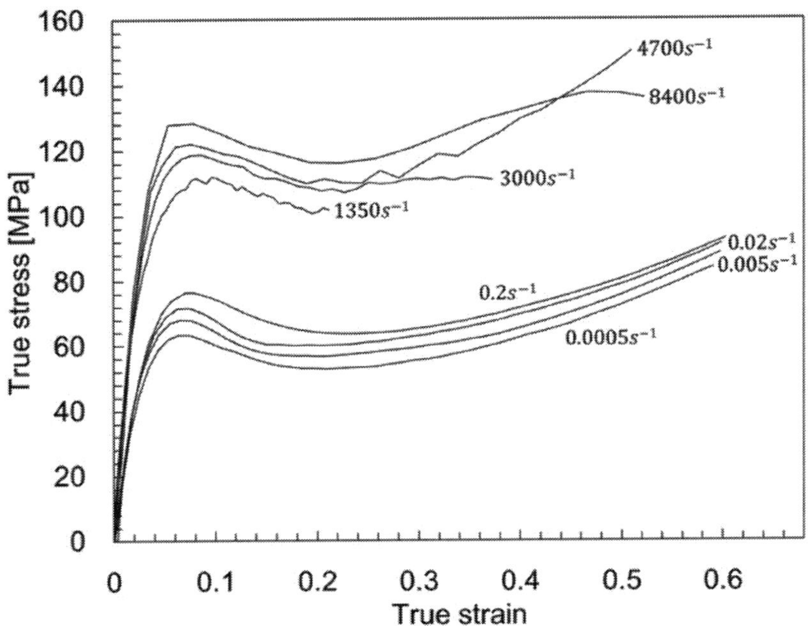

Figure 1.9. *True stress–strain response of polycarbonate under uniaxial compression loading at different strain rates*

Figure 1.10 shows dependences of yield stress and yield drop on the strain rate. It should be noted that yield drop is defined as the difference between the upper yield stress and the lower yield stress. In quasi-static testing (low strain rates), both upper and lower yield stresses linearly increase with the logarithmic strain rate. These linear relations, i.e. the so-called bilinear relations, can also be observed in dynamic testing (high strain rates) with a large slope and are strongly in accordance with the results of many studies (Boyce *et al.* 1994; Richeton *et al.* 2006). Mulliken and Boyce (2006) indicated the existence of a significant material transition point when the strain rate is about 150 s^{-1} for polycarbonate. In our study, the transition point obtained experimentally takes place at the strain rate of 132 s^{-1}. As mentioned above, the strain softening behavior of polycarbonate material is represented by yield drop, which remains constant and is therefore independent of the strain rate.

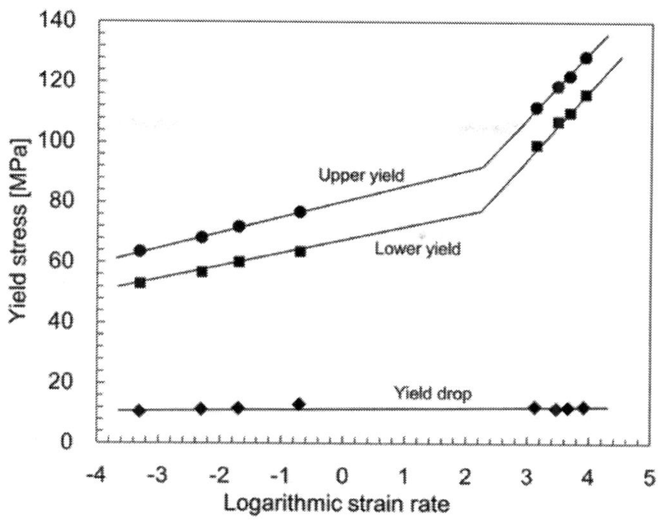

Figure 1.10. *Yield stress and yield drop versus logarithmic strain rate*

1.2. Uniaxial tension tests at various strain rates

1.2.1. *Experimental setup of quasi-static uniaxial tension tests*

Dog-bone shaped specimens are recommended for uniaxial tension tests (ASTM D638 2003) in order to reduce the influence of stress concentrations induced by loading grips. The details of recommended specimen dimensions are provided in ASTM D638 (2003). In the present test, specimens are injection-molded with the dimensions shown in Figure 1.11.

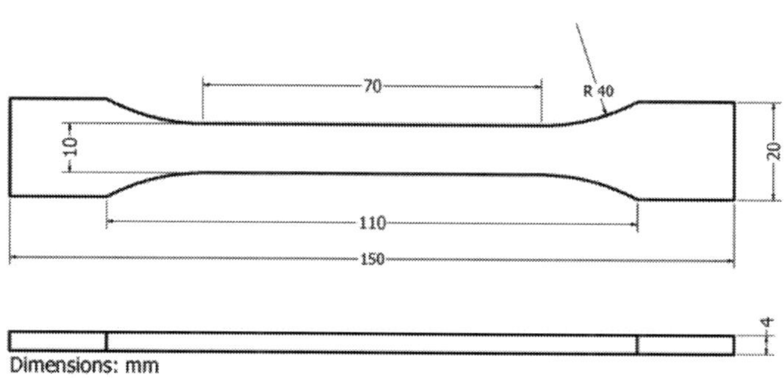

Figure 1.11. *Dimensions of the polycarbonate specimen for the quasi-static uniaxial tension test*

Quasi-static uniaxial tension tests are performed on a TESTRESOURCES 310R electromechanical universal testing system. The schematic is shown in Figure 1.12. The specimen is fixed in two wedge-type grips to restrain axial and rotational motions. The lower grip is fixed to the machine base, while the other steadily moves with the cross head at a constant loading rate. The force–displacement results are then converted into the true stress–strain response using equations [1.1]–[1.6].

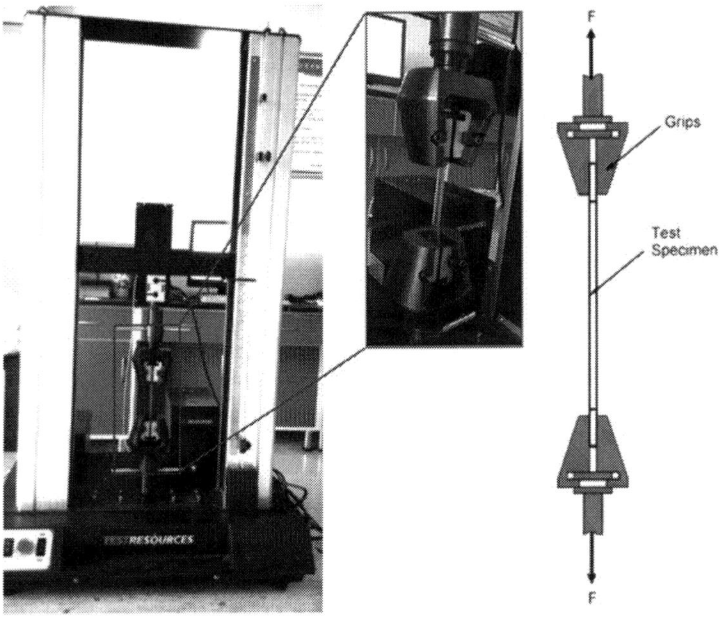

Figure 1.12. *Experimental setup of the quasi-static compression test*

1.2.2. *Experimental setup of dynamic uniaxial tension tests*

The setup change of the SHPB into a split Hopkinson tension bar (SHTB) is needed for the dynamic uniaxial tension test. The specimen is usually threaded or bonded between the incident and transmission bars. A tensile pulse is directly generated in the incident bar by incorporating a flange on the free end of the incident bar and a striker tube surrounding the incident bar (Yin and Wang 2010; Xu *et al.* 2016). The striker bar is then propelled towards the flange away from the specimen to induce the incident tensile pulse, as shown in Figure 1.13.

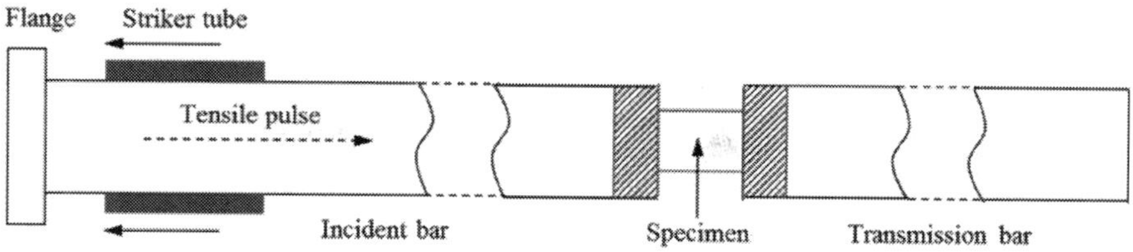

Figure 1.13. *Illustration of using a flange and a striker tube for generating tensile pulse. For a color version of this figure, see www.iste.co.uk/zhang/polycarbonate.zip*

An alternative method is to use a split collar to transmit the initial compressive pulse into tensile pulse (Sarva and Boyce 2007). As shown in Figure 1.14, a split collar surrounds the specimen and is snugly sandwiched between the two bars. This collar enables the transmission of the initial compressive pulse into the second bar without loading the specimen. The subsequent reflected tensile pulse from the free rear end of the second bar then loads the specimen.

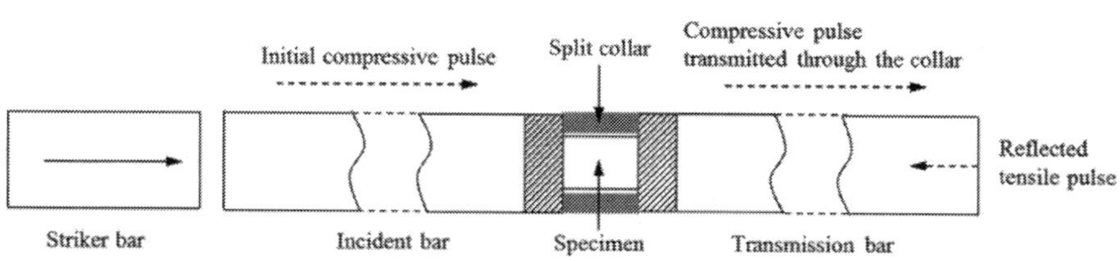

Figure 1.14. *Illustration of using a split collar for generating tensile pulse. For a color version of this figure, see www.iste.co.uk/zhang/polycarbonate.zip*

In this study, the SHTB incorporating a flange and a striker tube is used to perform dynamic tension testing on polycarbonates. Dumbbell-shaped specimens are used. As with the SHPB test, a short gauge length is suggested to facilitate early dynamic equilibrium in the specimen. The dimensions of the specimen used are shown in Figure 1.15. It should be noted that right-hand and left-hand threads are connected with the incident and transmission bars. All the specimens are carefully manufactured from polycarbonate rods subjected to the extrusion process and stored in a dry atmosphere at room temperature (25°C) to release residual stress caused by the manufacturing process.

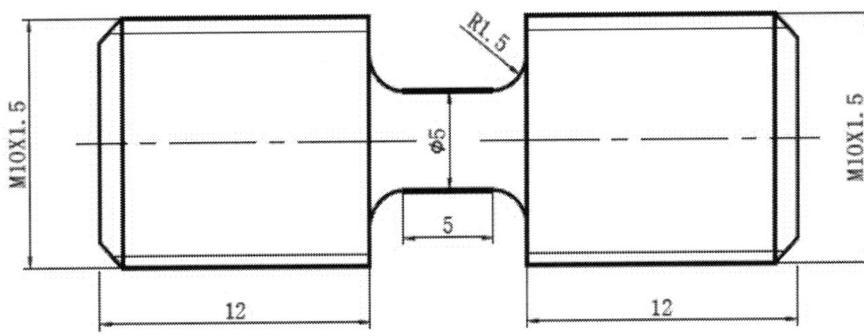

Figure 1.15. *Dimensions of the polycarbonate specimen for the SHTB test*

Figure 1.16 shows the schematic of the SHTB setup that consists of an air gun system, a striker tube, an incident bar with a transfer flange at one end, a transmission bar, a momentum trap bar and the data acquisition system. The striker tube and all the bars are made of 18Ni steel and behave elastically during the test. The striker tube is 380 mm in length with an outer diameter of 26 mm and an inner diameter of 19 mm. The incident and transmission tension bars are 2,800 and 1,200 mm in length respectively, and both of them have a diameter of 19 mm. The momentum trap bar is 430 mm in length and 19 mm in diameter. Both the incident and transmission bars contain threaded holes to connect the specimen. The air gun system is used to propel the striker tube. A detailed illustration of the working principle of the air gun system can be found in Xu *et al.* (2012).

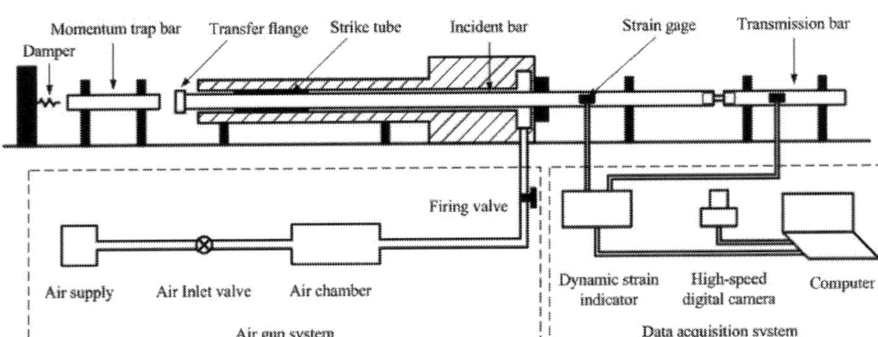

(a) Schematic diagram

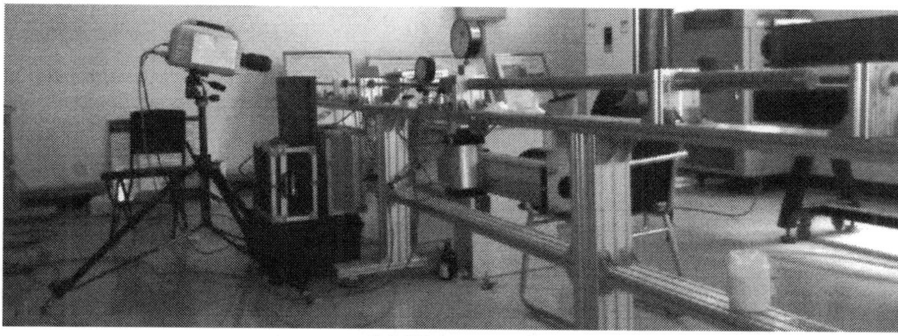

(b) Photograph

Figure 1.16. *SHTB setup*

The striker tube, accelerated by the air gun, slides along the incident bar which impacts the flange to generate a tensile pulse in the incident bar. The duration of the tensile pulse can be controlled by adjusting the length of the striker tube. As shown in Figure 1.16(a), a gap is precisely preset to separate the momentum trap bar and the transfer flange of the incident bar. The end of the momentum trap bar and the face of the transfer flange are brought in contact once the tensile pulse generated by the striker tube is completely transferred into the incident bar through the transfer flange. The generated tensile pulse travels along the incident bar and then propagates into the specimen. The stress wave travels along the specimen and creates tension deformation within the specimen. Following this, the stress wave is partly transmitted into the transmission bar and partly reflected as a compressive pulse back into the incident bar. The reflected compressive pulse is then transmitted into the momentum trap bar and reflects off the free end of this bar as a tensile pulse. Since the contact interface with the transfer flange cannot support tension, this tensile pulse is trapped in the momentum trap bar. Therefore, the reloading of the specimen by the reflected pulse is effectively prevented.

The theoretical value of the separation between the momentum trap bar and the transfer flange, u_s, can be estimated as (Xu *et al.* 2012)

$$u_s = 2C_b \int_0^t \varepsilon_f(t)\,dt \qquad [1.16]$$

where C_b is the wave propagation speed in the bar, which can be determined using equation 7, and $\varepsilon_f(t)$ is the compression strain in the flange. In practice, the calculated u_s will be further optimized by a few trials.

For SHTB tests, the same principles of data analysis as described for SHPB tests in section 1.1.2 can be adopted.

1.2.3. *Experimental results*

Quasi-static uniaxial tension tests are carried out at strain rates of 0.5×10^{-3}, 1.0×10^{-3}, 1.0×10^{-2} and 1.0×10^{-1} s^{-1}. The true stress–strain responses of polycarbonate are shown in Figure 1.17. Roughly speaking, the tensile stress–strain responses, including initial elasticity, yield transition, strain softening and strain hardening, are similar to the compressive ones. The yield stress and flow stress slightly increase with the increase in the strain rate.

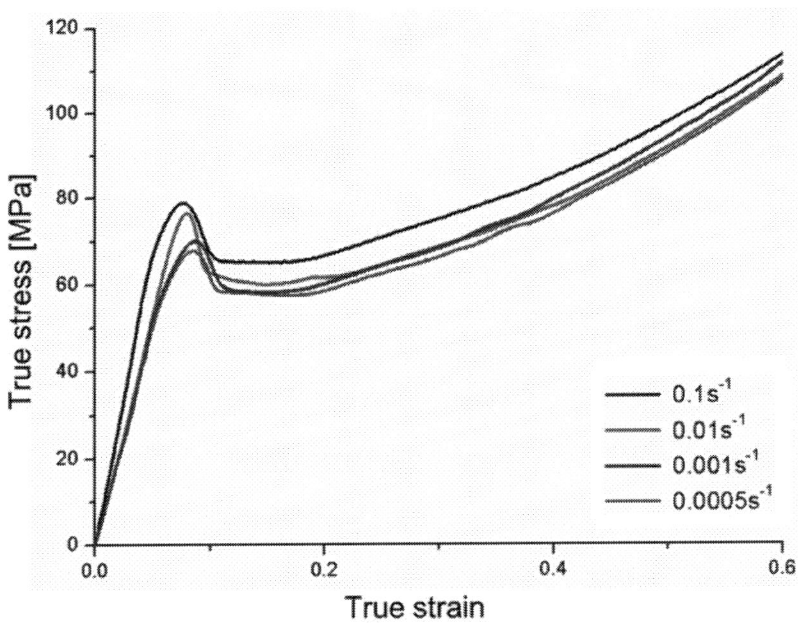

Figure 1.17. *True stress–strain responses of polycarbonate under quasi-static uniaxial tension loadings. For a color version of this figure, see www.iste.co.uk/zhang/polycarbonate.zip*

However, considerable differences exist for the yielding point and strain hardening in tension and compression. Figure 1.18 shows the true stress–strain responses at the strain rate of 0.1 s^{-1} for both uniaxial tension and compression. It is observed that the yield stress in tension is lower than that in compression due to the pressure sensitivity of yielding (Spitzig and Richmond 1979; Caddell and Kim 1981). Moreover, the strain hardening in tension occurs at low strains. The underlying mechanism is that molecular chains in tension are aligned along the elongation direction against loading, whereas compression forces the chains to align in a plane normal to the compression direction.

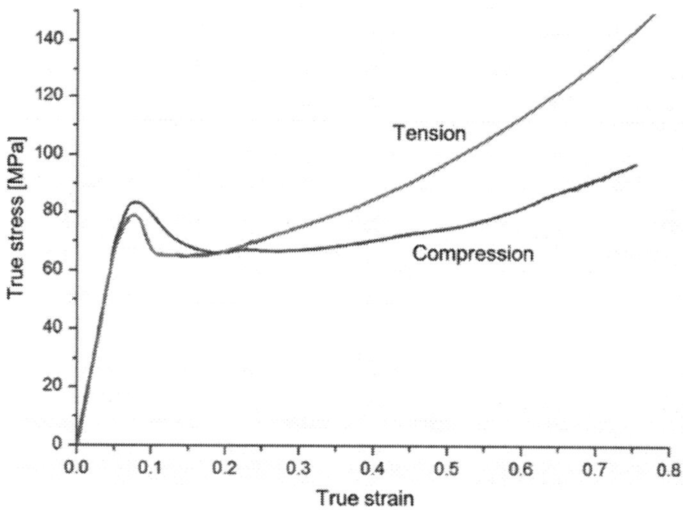

Figure 1.18. *True stress–strain responses of polycarbonate in uniaxial tension and compression at the strain rate of 0.1 s^{-1}. For a color version of this figure, see www.iste.co.uk/zhang/polycarbonate.zip*

At the strain rate of 1,400 s^{-1}, strain signals at the incident and transmission bars are extracted from the strain acquisition system, as shown in Figure 1.19. Obviously, a nearly flat plateau exists in the reflected pulse, which means a steady strain rate condition.

Dynamic tension tests are further carried out at different strain rates. The strain rate histories are determined using equation [1.15], and the results are given in Figure 1.20. The flat portions of strain rates confirm that almost constant strain rate loadings are achieved in all the tests.

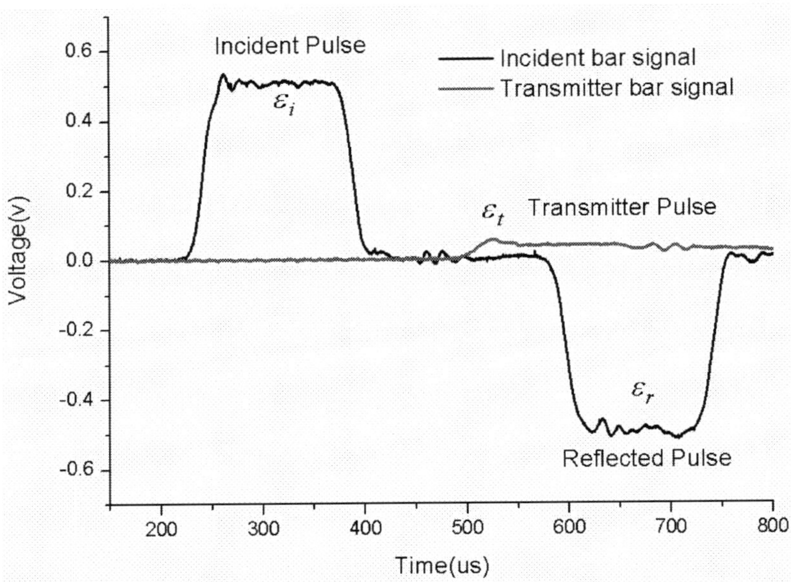

Figure 1.19. *Typical signals measured from the gauges on the bars. For a color version of this figure, see www.iste.co.uk/zhang/polycarbonate.zip*

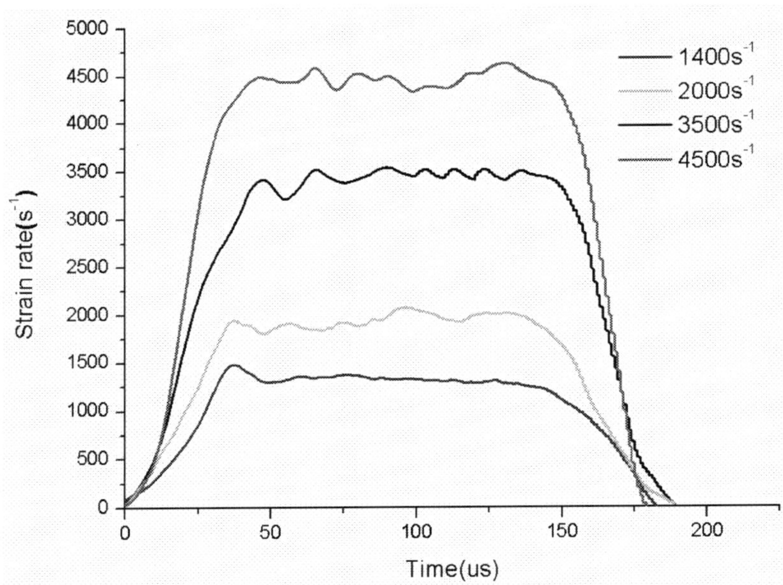

Figure 1.20. *Strain rate histories of the specimens in all the tests. For a color version of this figure, see www.iste.co.uk/zhang/polycarbonate.zip*

Figure 1.21 presents the true stress–strain responses of polycarbonate at various strain rates under dynamic tension loadings. Compared with Figure 1.17, more fluctuations exist in the dynamic tensile stress–strain curves due to the imperceptible gaps in the threaded connections of the bars and the specimen. This phenomenon is in agreement with the work of Sarva and Boyce (2007). The dynamic response of polycarbonate is significantly distinct from the quasi-static one. In the case of quasi-static loading, polycarbonate specimens are deformed in a ductile manner up to a strain of 0.6. Furthermore, the yield stress and flow stress slightly increase with the increase in the strain rate. In the case of dynamic loading, yield stress is apparently higher than that under quasi-static loading.

Here, it should be mentioned that the hardening behavior is not captured at strain rates of 1,400 and 2,000 s^{-1} because the duration of the incident stress pulse is restricted by the length of the prefixed steel bar. Since the duration of the incident stress pulse is constant in all the high rate tests, the total measurable uniaxial strain decreases with the reduction of the strain rate. Therefore, for strain rates of 1,400 and 2,000 s^{-1}, only the softening behavior is recorded.

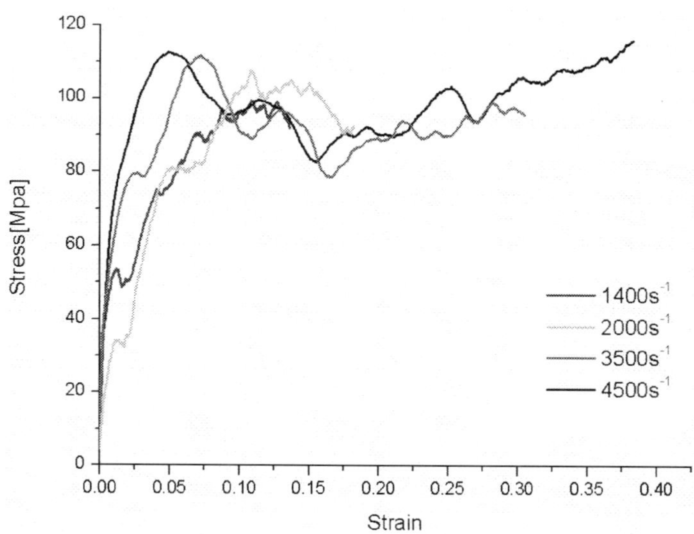

Figure 1.21. *True stress–strain in uniaxial tension at different strain rates. For a color version of this figure, see www.iste.co.uk/zhang/polycarbonate.zip*

Figure 1.22 shows the yield stress and strain softening of polycarbonate from quasi-static loading to dynamic loading. Likewise, strain softening demonstrates a dramatic stress drop after yielding. This phenomenon is considered as a degradation of material caused by the rupture of the molecular chain network. Although both upper and lower yield stresses increase in bilinear relations versus the logarithmic strain rate, the yield drop is an intrinsic nature of amorphous polymers and remains constant. This indicates that strain softening in tension is also independent of the strain rate as in the compression test. Finally, Table 1.1 summarizes experimental setups, mechanical behaviors and the geometry of the specific specimen in uniaxial compression and tension tests.

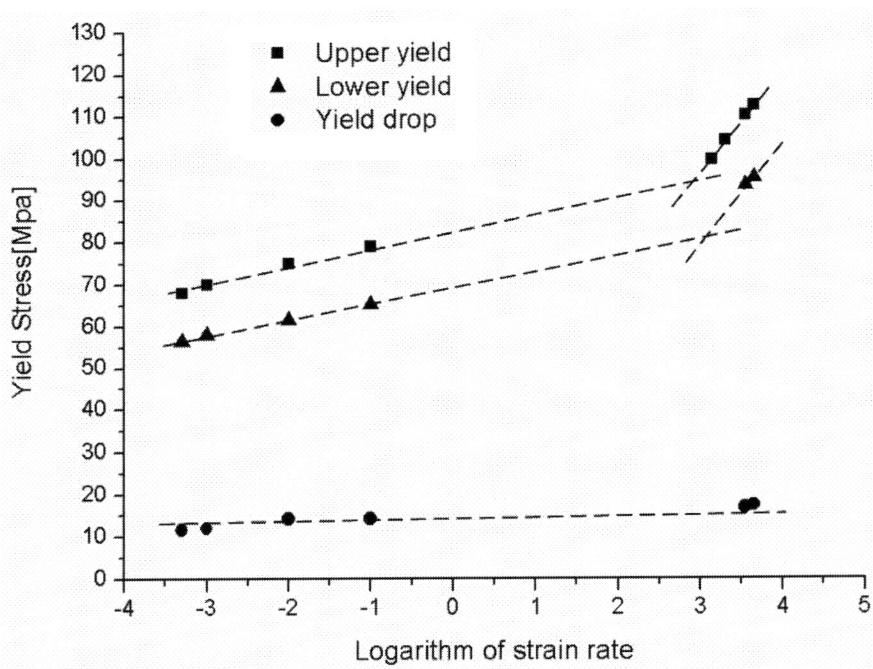

Figure 1.22. *Yield stress and yield drop versus logarithmic strain rate in tension*

<table>
<tr><th></th><th>Uniaxial compression test</th><th>Uniaxial tension test</th></tr>
<tr><td>Quasi-static test specimen</td><td>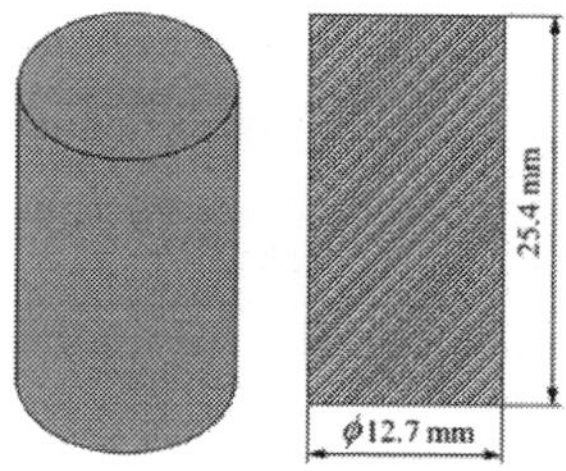
</td><td>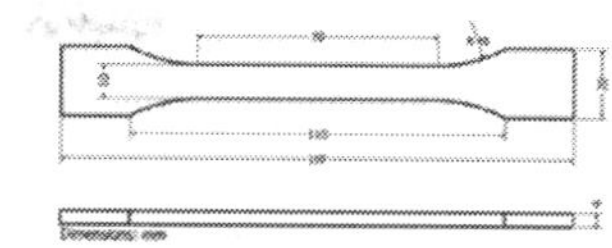</td></tr>
<tr><td>Quasi-static test setup</td><td>Universal testing system
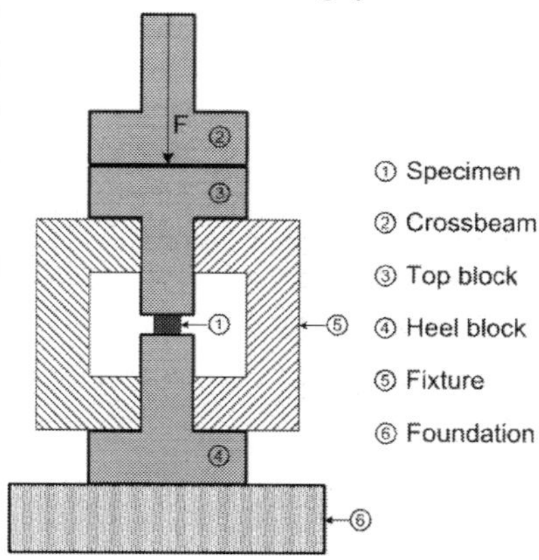
</td><td>Universal testing system
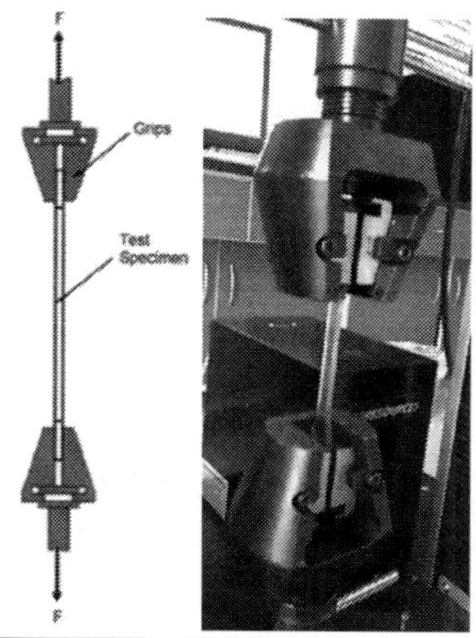
</td></tr>
<tr><td>Dynamic test specimen</td><td>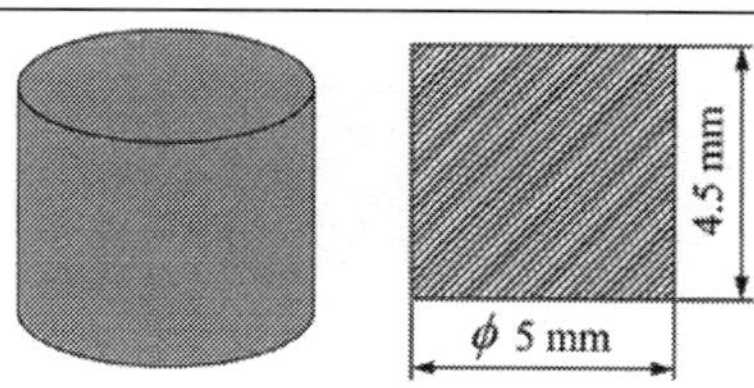
</td><td>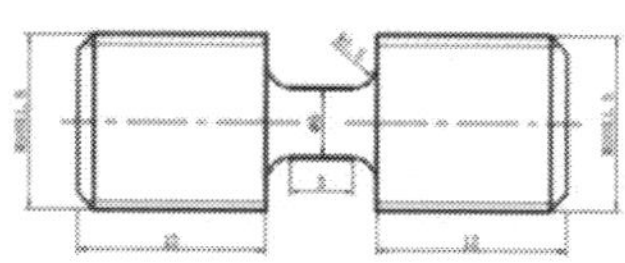</td></tr>
<tr><td>Dynamic test setup</td><td>SHPB
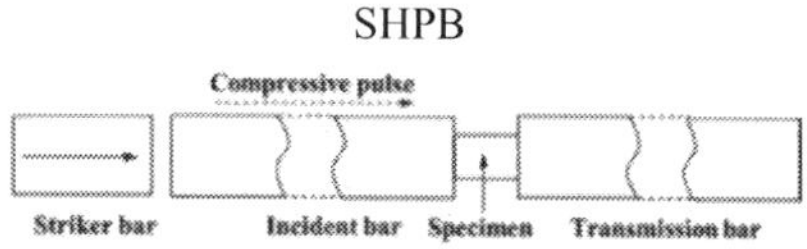
</td><td>SHTB
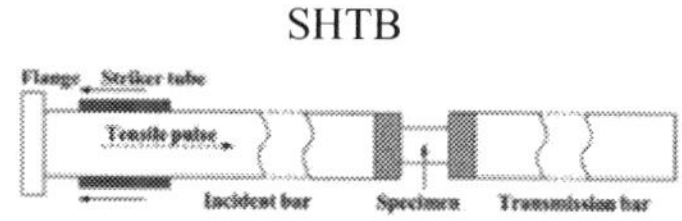
</td></tr>
<tr><td>Mechanical behavior</td><td colspan="2">(1) Stress–strain response shows nonlinear behavior characterized by elasticity, yield transition, strain softening and strain hardening.
(2) Both upper and lower yield stresses increase in bilinear relations versus the logarithmic strain rate while the yield drop remains constant.
(3) The yield stress in tension is lower than that in compression and the strain hardening in tension occurs at lower strains.</td></tr>
</table>

Table 1.1. *Experimental setups, mechanical behaviors and the geometry of the specific polycarbonate specimen in uniaxial compression and tension tests*

1.3. Quasi-static uniaxial compression tests at various temperatures

This section is focused on the effects of temperature on the mechanical properties of polycarbonate. Here, only the quasi-static uniaxial compression tests are carried out. The specimen geometry is the same as that used for room-temperature tests. Quasi-static uniaxial compression tests are conducted on a CRIMS DNS 100 electromechanical universal testing system over a range of temperature from -60 to 120°C at a constant strain rate of $5 \times 10^{-4}\ s^{-1}$.

Figure 1.23. *Temperature chamber attached to the compression platform*

A temperature regulating chamber is attached to the compression platform, as shown in Figure 1.23. The temperature enclosure can provide a temperature up to 120°C using electric heaters and low cryogenic temperature up to -60°C using a liquid nitrogen-cooled environment. During testing, temperatures are monitored through a thermocouple and specimens are preheated at testing temperatures for 5 minutes.

The true stress–strain responses of polycarbonate under uniaxial compression loading at various temperatures are shown in Figure 1.24. With the increase in temperature from -60 to 120°C, an obvious downward shift in stress–strain responses is observed. Specifically, the yield stress and the plastic flow stress decrease simultaneously. In this study, the glass transition temperature of the specimens is about 148°C. When the testing temperature approaches this value, the polycarbonate material becomes weak and soft. The yield stress is lower than 40 MPa, and the strain hardening behavior almost vanishes.

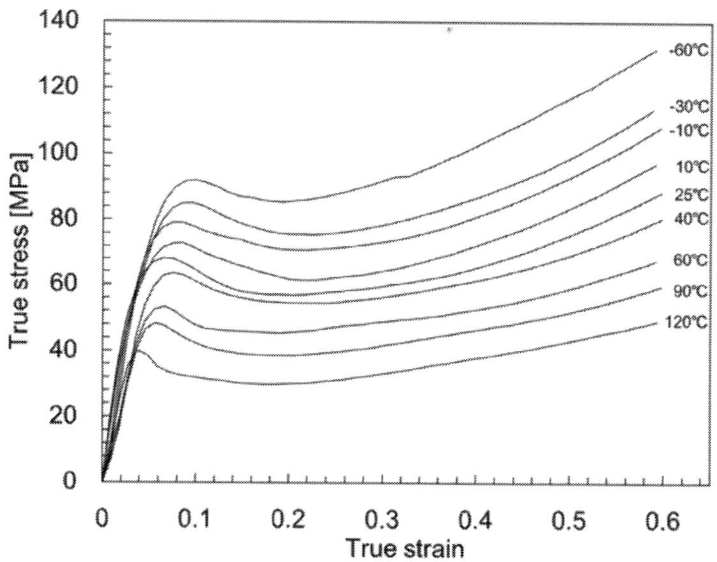

Figure 1.24. *True stress–strain response of polycarbonate under uniaxial compression loading at various temperatures*

According to previous studies (Brooks *et al.* 1998; Richeton *et al.* 2006), the yield stress of polymers depends on the glass (α) transition and secondary (β) transition temperatures. Yield stress is found to linearly decrease between the α and β transition temperatures. Here, similar experimental results can be observed. Figure 1.25 shows the yield stress and the yield drop versus the temperature. Both the upper and lower yield stresses linearly decrease with the temperature. The yield drop remains nearly constant over the temperature range, indicating that the strain softening of polycarbonate is independent of temperature.

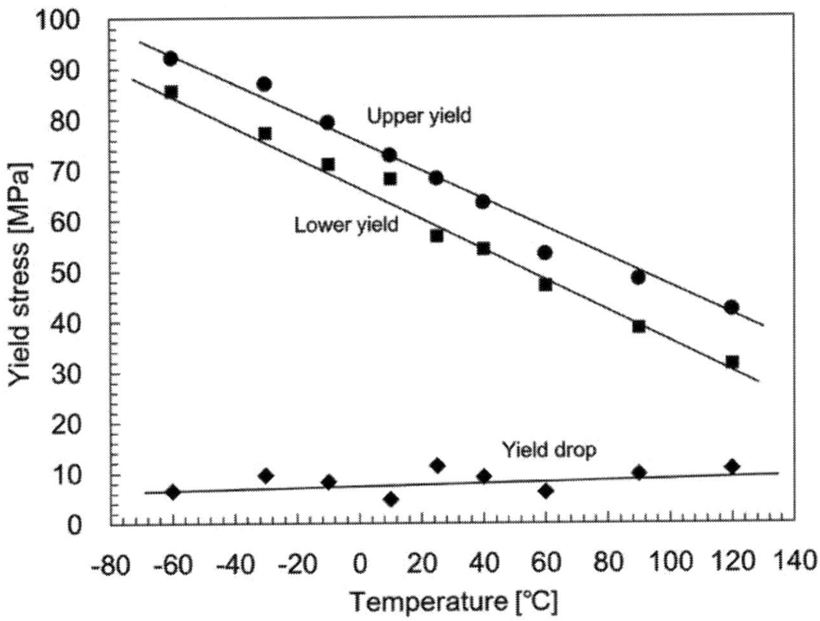

Figure 1.25. *Yield stress and yield drop versus temperature*

1.4. Conclusion

This chapter has focused on quasi-static and dynamic uniaxial tension and compression tests of polycarbonate over large ranges of strain rates and temperatures. The high strain rate compression and tension tests were carried out using the SHPB and SHTB respectively. At room temperature and strain rates ranging from 5×10^{-4} to 8.4×10^{3} s^{-1}, the true stress–strain responses were nonlinear and characterized by initial elasticity, yield transition, strain softening and strain hardening. Among others, the yield stress dramatically increased with the increase in the strain rate regardless of whether the loading test was in compression or in tension. Nevertheless, the yield stress in tension was lower than that in compression. The strain hardening in tension occurred at lower strains.

The quasi-static uniaxial compression tests were conducted at temperatures ranging from −60 to 120°C with a constant strain rate of 5×10^{-4} s^{-1}. With the increase in temperature, an obvious downward shift in stress–strain response was observed. The yield stress and the plastic flow stress decreased simultaneously. For both upper yield and lower yield, a linear decline with temperature was observed. It should be noted that the dynamic compression and tension tests were not addressed in the study of temperature effects. However, the temperature dependence law obtained from quasi-static compression tests could provide a guideline for polycarbonate under other loading conditions.

1.5. References

ASTM D638. (2003). Standard test method for tensile properties of plastics. *Book of Standards Volume 08.01*, ASTM International, Pennsylvania.

ASTM D695 (2015). Standard test method for compressive properties of rigid plastics. *Book of Standards Volume 08.*01, ASTM International, Pennsylvania.

Boyce, M.C., Arruda, E.M. and Jayachandran, R. (1994). The large strain compression, tension, and simple shear of polycarbonate. *Polymer Engineering and Science*, 34, 716–725.

Brooks, N.W.J., Duckett, R.A. and Ward IM. (1998). Temperature and strain-rate dependence of yield stress of polyethylene. *Journal of Polymer Science Part B*, 36, 2177–2189.

Caddell, R.M. and Kim, J.W. (1981). Influence of hydrostatic pressure on the yield strength of anisotropic polycarbonate. *International Journal of Mechanical Sciences*, 23, 99–104.

Chen, W., Lu, F., Frew, D.J. and Forrestal, M.J. (2002). Dynamic compression testing of soft materials. *Journal of Applied Mechanics*, 69, 214–223.

Chen, W. and Song, B. (2011). *Split Hopkinson (Kolsky) Bar: Design, Testing and Applications*. Springer, New York.

Gray, G.T. (1985a). Classic split-Hopkinson pressure bar testing. *ASM Handbook Volume 8: Mechanical Testing and Evaluation*, ASM International.

Gray, G.T. (1985b). High strain rate testing. *ASM Handbook Volume 8: Mechanical Testing and Evaluation*, ASM International.

Kolsky, H. (1963). *Stress Waves in Solids*. Dover, New York.

Mulliken, A.D. and Boyce, M.C. (2006). Mechanics of the rate-dependent elastic-plastic deformation of glassy polymers from low to high strain rates. *International Journal of Solids and Structures*, 43, 1331–1356.

Richeton, J., Ahzi, S., Vecchio, K.S., Jiang, F.C. and Adharapurapu, R.R. (2006). Influence of temperature and strain rate on the mechanical behavior of three amorphous polymers: characterization and modeling of the compressive yield stress. *International Journal of Solids and* Structures, 43, 2318–2335.

Sarva, S.S. and Boyce, M.C. (2007). Mechanics of polycarbonate during high-rate tension. *Journal of Mechanics of Materials and Structures*, 2, 1853–1880.

Spitzig, W.A. and Richmond, O. (1979). Effect of hydrostatic pressure on the deformation behavior of polyethylene and polycarbonate in tension and in compression. *Polymer Engineering and Science*, 19, 1129–1139.

Wang, L., Labibes, K., Azari, Z. and Pluvinage G. (1994). Generalization of split Hopkinson bar technique to use viscoelastic bars. *International Journal of Impact Engineering*, 15, 669–686.

Xu, Y.J., Gao, T.L. and Zhang, W.H. (2016). Experimentation and modeling of the tension behavior of polycarbonate at high strain rates. *Polymers*, 8(3), 63.

Xu, Z., Li, Y. and Huang, F. (2012). Application of split Hopkinson tension bar technique to study of dynamic fracture properties of materials. *Acta Mechanica Sinica*, 28, 424–431.

Yin, Z.N. and Wang, T.J. (2010). Deformation response and constitutive modeling of PC, ABS and PC/ABS alloys under impact tensile loading. *Materials Science and Engineering A*, 527, 1461–1468.

2

Constitutive Models of Polycarbonate

A constitutive model is a methodical representation of two or more physical quantities that describe material behavior under different loading conditions. In solid mechanics, the constitutive model generally describes the mechanical behavior of the material in terms of stress–strain or force–deformation relations. In this book, discussions about constitutive models are limited to the scope of solid mechanics.

An appropriate constitutive model obtained from elementary experiments can be used beyond the experimental conditions. Once a constitutive model is established, the equilibrium equation of any complex structure can be derived and further solved in terms of structural displacements or internal forces using the finite element method (FEM). Thus, the constitutive model is the foundation in structure analysis and design.

In Chapter 1, it can be seen that responses of polycarbonate depend on the strain rate and the temperature. Polycarbonate exhibits obvious elastic–viscoplastic behavior, including linear elasticity, yield transition, strain softening and strain hardening under compressive and tensile loading.

This chapter first gives a brief introduction to the existing constitutive models of polycarbonates. Some of the commonly used models from a simple linear viscoelastic model to a nonlinear viscoplastic model are presented. A phenomenological elastic–viscoplastic model is then developed for polycarbonate based on the experimental results presented in Chapter 1. The fundamental principle and approach to modeling complicated mechanical behavior over a wide range of strain rates and temperatures are discussed. An algorithmic scheme is presented for the numerical implementation of the proposed model.

2.1. Introduction to constitutive models for polycarbonate

In general, two kinds of constitutive models, i.e. microscopic and macroscopic models, are available in the literature. Quantum, molecular and atomistic theories are often used in microscopic constitutive modeling to characterize discontinuous atomic structures with molecules and large gaps between them. Such theories are computationally expensive for solving engineering problems and are not considered in this chapter.

The macroscopic model is often established within the framework of continuum mechanics based on elementary tests. Section 2.1.1 presents the simple linear viscoelastic model. Section 2.1.2 summarizes some of the most commonly used viscoplastic models and the developing trends in the constitutive modeling of polycarbonate.

2.1.1. *Linear viscoelastic model*

Polycarbonate exhibits a strain rate-dependent response when undergoing a small deformation. It is simplest to use the linear viscoelastic model for the description.

A linear viscoelastic model can be described using the rheological representation, which gives a direct illustration of viscous and elastic responses in the uniaxial stress condition. The rheological representation consists of springs (with elastic moduli) and dashpots (with viscous coefficients) (Christensen 2012; Flügge 2013). The standard spring and dashpot elements are shown in Figure 2.1. The stress–strain relations in spring and dashpot can be written as

$$\begin{aligned} \varepsilon &= \frac{1}{E}\sigma \quad \text{for spring} \\ \dot{\varepsilon} &= \frac{1}{\eta}\sigma \quad \text{for dashpot} \end{aligned} \qquad [2.1]$$

where σ is the stress, ε and $\dot{\varepsilon}$ represent the strain and the strain rate respectively, E is the elastic modulus and η is the viscous coefficient.

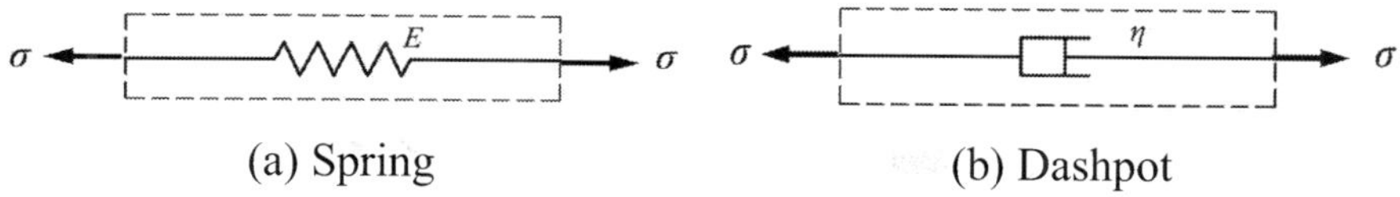

Figure 2.1. *Elements used in constructing viscoelastic models*

Spring and dashpot elements can be combined to yield different viscoelastic models. The simplest viscoelastic models are those named after Maxwell and Kelvin.

Maxwell model

In the Maxwell model, a spring with the elastic modulus E and a dashpot with the viscous coefficient η are connected in series, as shown in Figure 2.2.

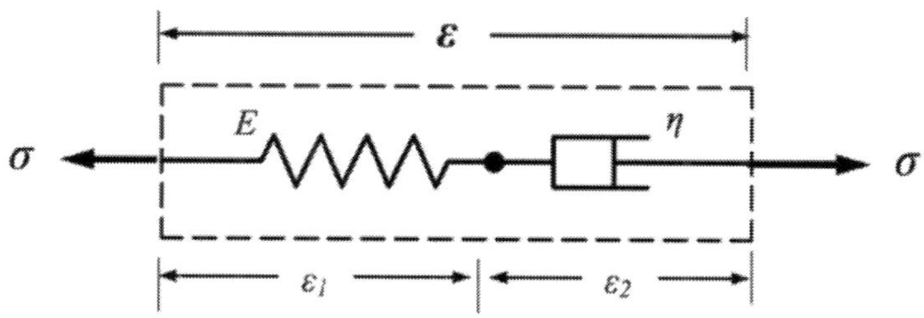

Figure 2.2. *Rheological representation of the Maxwell model*

Since stress in the spring is the same as that in the dashpot, the total strain in the system can be obtained by summing the spring strain (ε_1) and the dashpot strain (ε_2):

$$\varepsilon = \varepsilon_1 + \varepsilon_2 \qquad [2.2]$$

Differentiation of equation [2.2] gives rise to the stress–strain relation of the system

$$\dot{\varepsilon} = \frac{\partial}{\partial t}(\varepsilon_1 + \varepsilon_2) = \frac{\dot{\sigma}}{E} + \frac{\sigma}{\eta} \qquad [2.3]$$

where the symbol $\partial/\partial t$ is a differential operator. At a constant strain rate, Figure 2.3 shows the stress–strain relation of the Maxwell model.

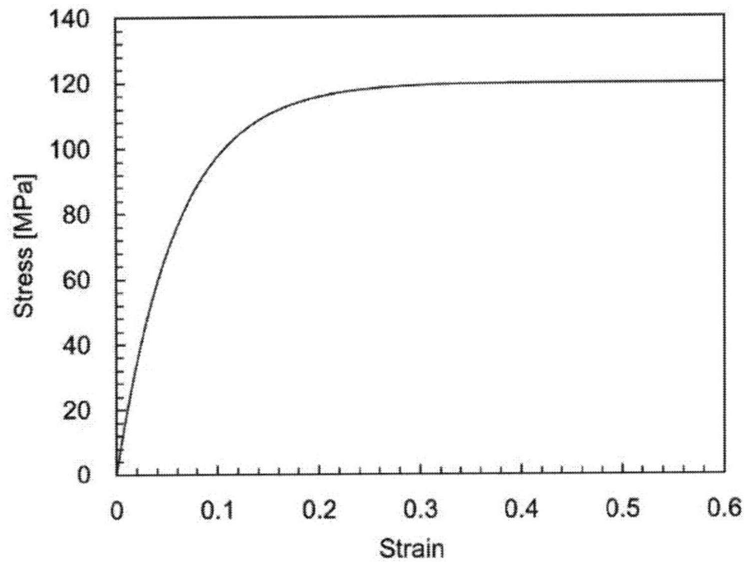

Figure 2.3. *Stress–strain relation of Maxwell model*

Kelvin model

The Kelvin or Voigt model is constructed by a parallel connection of the spring and dashpot, as shown in Figure 2.4.

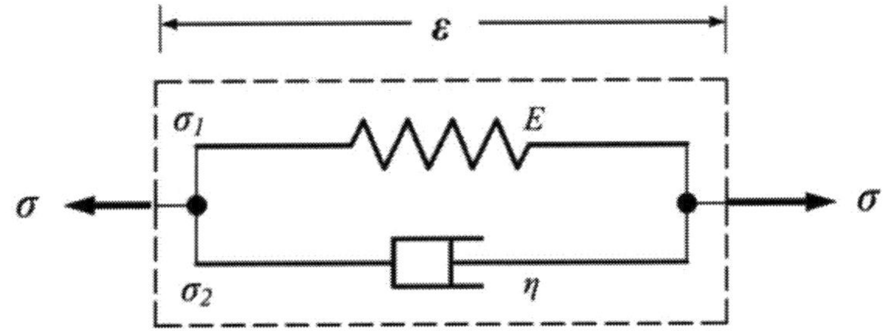

Figure 2.4. *Rheological representation of the Kelvin model*

The strain experienced by the spring is the same as that experienced by the dashpot. The total stress in the system can be obtained by summing the stresses in the spring and dashpot:

$$\sigma = \sigma_1 + \sigma_2 \qquad [2.4]$$

The stress–strain relation of the system can be given by

$$\sigma = E\varepsilon + \eta\dot{\varepsilon} \qquad [2.5]$$

Figure 2.5 shows the stress–strain relation of the Kelvin model at a constant strain rate.

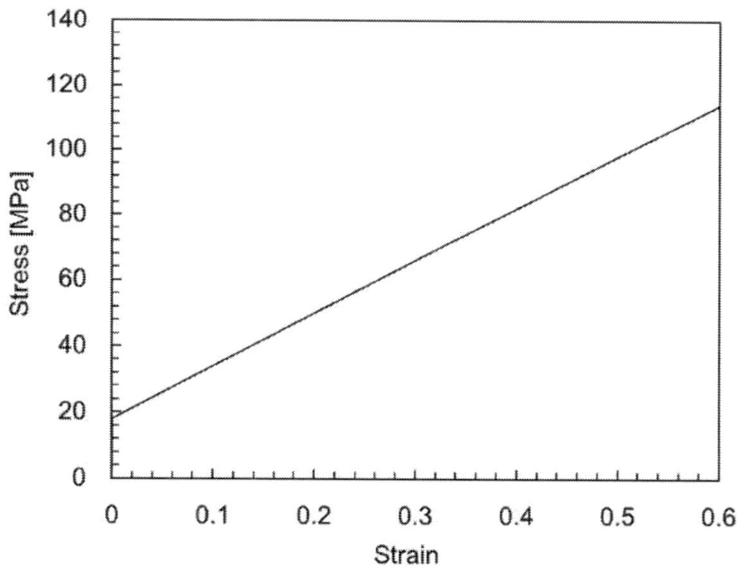

Figure 2.5. *Stress–strain relation of the Kelvin model*

Generalized Maxwell and Kelvin models

Maxwell and Kelvin models are the simplest viscoelastic models. Undoubtedly, a more accurate viscoelastic response can be modeled by using more spring and dashpot elements. The generalized model can be constructed by using various combinations of spring and dashpot elements in serial or parallel arrangement. However, more elements will increase the number of model parameters and sequentially increase the complexity of the model.

For example, Figure 2.6 shows the generalized Maxwell model consisting of a free spring and *n* Maxwell units in parallel.

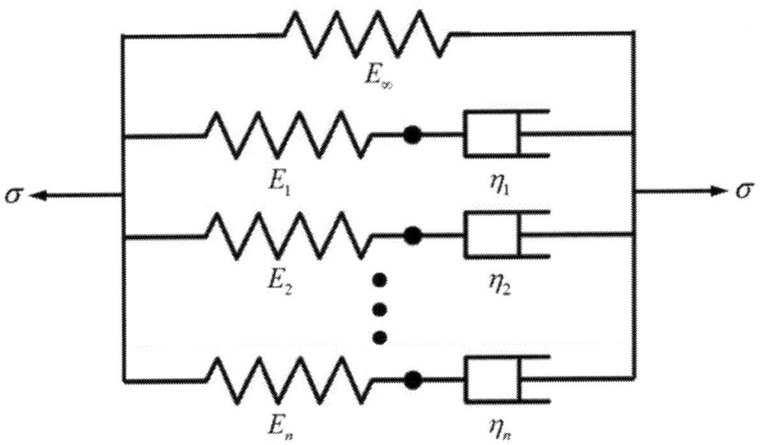

Figure 2.6. *Rheological representation of the generalized Maxwell model*

Assume that the elastic moduli of the free spring and the *i*th spring element are E_∞ and E_i respectively. The viscous coefficient of the *i*th dashpot element is denoted by η_i. Based on equation [2.3], the stress and strain σ_i and ε_i respectively in each Maxwell unit hold the relation

$$\frac{\partial \varepsilon_i}{\partial t} = \frac{1}{E_i}\frac{\partial \sigma_i}{\partial t} + \frac{\sigma_i}{\eta_i} \qquad [2.6]$$

Let ε be the external strain imposed on the whole model and σ be the total stress. The following relations then hold:

$$\begin{aligned} &\varepsilon = \varepsilon_\infty = \varepsilon_i \\ &\sigma = \sigma_\infty + \sum_{i=1}^{n} \sigma_i \end{aligned} \qquad [2.7]$$

where ε_∞ and σ_∞ are the strain and stress of the free spring respectively. Figure 2.7 shows the stress–strain relation of the generalized Maxwell model, with n = 1, 2, 3, 4 and 5.

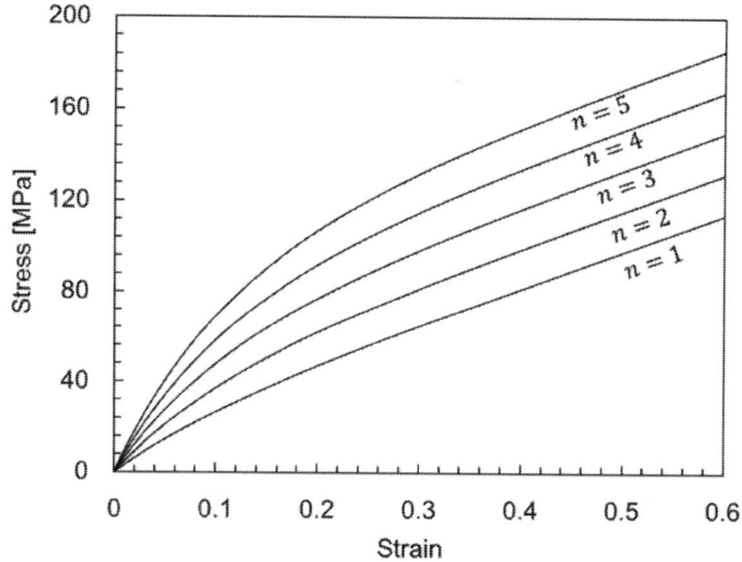

Figure 2.7. *Stress–strain relation of the generalized Maxwell model*

In the generalized Kelvin model, a free spring and n Kelvin units are connected in series, as shown in Figure 2.8.

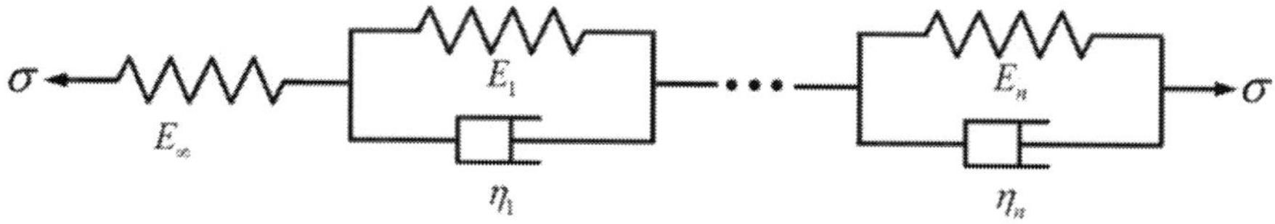

Figure 2.8. *Rheological representation of the generalized Kelvin model*

Based on equation [2.5], the stress and strain σ_i and ε_i respectively in each Kelvin unit are related by

$$\sigma_i = E_i \varepsilon_i + \eta_i \frac{\partial \varepsilon_i}{\partial t} \qquad [2.8]$$

The stress at each unit is identical, while the total strain is the sum of the internal strains of all units. Thus, the total stress and strain are expressed as

$$\sigma = \sigma_\infty = \sigma_i$$
$$\varepsilon = \varepsilon_\infty + \sum_{i=1}^{n} \varepsilon_i \qquad [2.9]$$

Figure 2.9 shows the stress–strain relation of the generalized Kelvin model, with $n = 1, 2, 3, 4$ and 5.

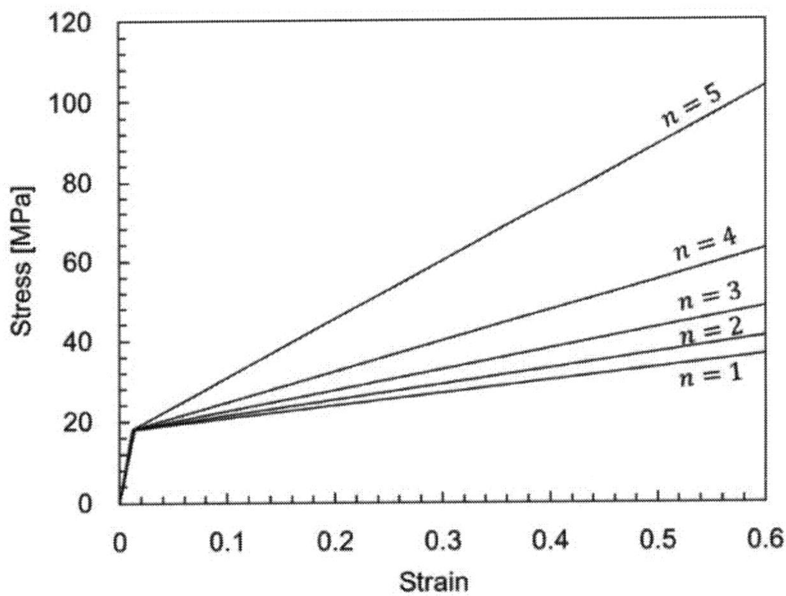

Figure 2.9. *Stress–strain relation of the generalized Kelvin model*

2.1.2. *Viscoplastic model*

The aforementioned viscoelastic models are limited to describing the viscoelastic deformation. In fact, polycarbonate exhibits complicated post-yield viscoplasticity, including strain softening, strain hardening as well as strain rate and temperature dependence. In order to capture these material characteristics, a viscoplastic model is supposed to comprise the following basic constitutive laws:

– a linear or nonlinear elastic law to define the relation between stress and elastic strain;

– a yield criterion to determine the viscoplastic regime start;

– a viscoplastic flow rule to control the flow direction and magnitude of the viscoplastic strain;

– a viscoplastic hardening law to determine the change in the yield surface.

Over the last decades, a large number of viscoplastic constitutive models have been proposed in the literature. These models can be grouped into two major categories: physically-based models and phenomenological models.

2.1.2.1. *Physically-based models*

The purpose of the physically-based models is to capture the viscoplastic behavior by including the mechanisms of microstructure evolution during deformation. Among these models, the “BPA” model, the “OGR” model and the “EGP” model are representative and commonly used.

BPA model

The BPA model was developed by Mary Boyce’s research group at MIT (Boyce *et al.* 1988; Arruda and Boyce 1993; Arruda *et al.* 1995) based on the mechanisms of molecular motions of polycarbonate. It assumes that the viscoplastic flow and hardening behavior are related to the chain segment rotation and the chain alignment. Both the intermolecular resistance to chain segment rotation and the entropic resistance to chain alignment contribute to the stress of polycarbonate during deformation.

Specifically, the rotation of chain segments of polycarbonate is composed of two parts: rotation of the main-chain segments (α process) and rotation of the main-chain phenyl group (β process). The stress induced by the intermolecular resistance can be expressed as (Argon 1973):

$$\boldsymbol{\sigma}_{A\alpha} = \frac{1}{J_\alpha}\phi_\alpha^e\left[\ln \mathbf{V}_{A\alpha}^e\right]$$
$$\boldsymbol{\sigma}_{A\beta} = \frac{1}{J_\beta}\phi_\beta^e\left[\ln \mathbf{V}_{A\beta}^e\right] \quad [2.10]$$

where $\boldsymbol{\sigma}_{Ai}\,(i=\alpha,\beta)$ represents the Cauchy stress induced by the intermolecular resistance to the rotations of the main-chain segments and the main-chain phenyl group; J_i is the corresponding elastic volume change; ϕ_i^e is the fourth-order modulus tensor; and $\ln \mathbf{V}_{Ai}^e$ is the Hencky strain.

The stress induced by the entropic resistance to chain alignment is defined using the Arruda–Boyce eight-chain model (Wang and Guth 1952):

$$\boldsymbol{\sigma}_B = \frac{C_R}{3}\frac{\sqrt{N}}{\lambda_{chain}^p}L^{-1}\left(\frac{\lambda_{chain}^p}{\sqrt{N}}\right)\overline{\mathbf{B}}' \quad [2.11]$$

where λ_{chain}^p is the stretch on a chain in the eight-chain network; C_R is the rubbery modulus; N is the number of rigid chain links between entanglements; L is the Langevin function defined by $L(\beta)=\coth\beta-\frac{1}{\beta}$; and $\overline{\mathbf{B}}'$ is the deviatoric part of the left Cauchy–Green tensor.

The total stress is given by the sum of the stresses induced by the intermolecular resistance to chain segment rotation and the entropic resistance to chain alignment:

$$\boldsymbol{\sigma} = \boldsymbol{\sigma}_{A\alpha} + \boldsymbol{\sigma}_{A\beta} + \boldsymbol{\sigma}_B \quad [2.12]$$

OGR model

The OGR model was developed by Paul Buckley's research group at Oxford University (Buckley and Jones 1995; Wu and Buckley 2004; De Focatiis *et al.* 2010). This model is based on the assumption that a deformed amorphous polymer stores strain energy by two mechanisms: perturbation of interatomic potentials (bond stretching) and perturbation of configurational entropy through the change in molecular conformations.

The Cauchy stress $\boldsymbol{\sigma}$ can be expressed as the sum of a bond stretching stress $\boldsymbol{\sigma}^b$ and a conformational stress $\boldsymbol{\sigma}^c$:

$$\boldsymbol{\sigma} = \boldsymbol{\sigma}^b + \boldsymbol{\sigma}^c \qquad [2.13]$$

The bond stretching stress is expressed as

$$\boldsymbol{\sigma}^b = \sum_{j=1}^{N} v_j \boldsymbol{\sigma}_j^b, \ \sum_{j=1}^{N} v_j = 1 \qquad [2.14.1]$$

$$\bar{\mathbf{D}} = \frac{\hat{\boldsymbol{\sigma}}_j^b}{2G_b} + \frac{\boldsymbol{\sigma}_j^b}{2G_b \tau_j} \qquad [2.14.2]$$

where v_j is the weighting of each bond-stretching stress; $\boldsymbol{\sigma}_j^b$ is the *j*th bond-stretching stress, which can be calculated using equation [2.14.2]; $\hat{\boldsymbol{\sigma}}_j^b$ is the objective rate of the *j*th bond-stretching stress; $\bar{\mathbf{D}}$ is the deviatoric rate of deformation; G_b is the bond-stretching shear modulus; and τ_j is the relaxation time associated with the *j*th bond-stretching stress.

The stresses $\boldsymbol{\sigma}_i^c$ from conformational entropy are obtained from a free-energy function ψ^c :

$$\boldsymbol{\sigma}_i^c = \bar{H}_i \frac{\partial \psi^c}{\partial \bar{H}_i} - p, \quad i = 1, 2, 3 \qquad [2.15]$$

where $\bar{H}_i$ are the eigenvalues of the deviatoric stretch $\bar{\mathbf{V}}$, p is an unknown pressure arising from the constraint $\det \bar{\mathbf{V}} = 1$ and ψ^c is given as

$$\psi^c = \frac{N_e k_B \theta}{2} \left[\frac{(1+\eta)(1-\alpha^2)}{(1-\alpha^2 \sum \lambda_i^2)} \sum \frac{\lambda_i^2}{1+\eta\lambda_i^2} + \sum ln\, 1 + \eta\lambda_i^2 + ln\left(1 - \alpha^2 \sum \lambda_i^2\right) \right] \qquad [2.16]$$

where N_e is the number of entanglements per unit volume, k_B is the Boltzmann constant, η is a dimensionless constant, α is a model parameter representing the inextensibility of the entangled network and λ_i $(i = 1{,}2{,}3)$ are the principle stretches.

EGP model

The EGP model was developed by Leon Govaert's research group in Eindhoven (Tervoort *et al.* 1997; Govaert *et al.* 2000; Klompen *et al.* 2005). The basis of this model is the split of the total stress into two parts

$$\boldsymbol{\sigma} = \boldsymbol{\sigma}_r + \boldsymbol{\sigma}_s \qquad [2.17]$$

where $\boldsymbol{\sigma}_r$ denotes the strain hardening contribution attributed to the molecular orientation of the entanglement network and $\boldsymbol{\sigma}_s$ represents the plastic flow contribution attributed to intermolecular interactions.

$\boldsymbol{\sigma}_r$ and $\boldsymbol{\sigma}_s$ are modeled using the neo-Hookean elastic expression and the nonlinear Maxwell expression respectively. The total stress is formulated using the following equations:

$$\begin{cases} \boldsymbol{\sigma} = H\widetilde{\boldsymbol{B}}^d + K(J-1)\boldsymbol{I} + G\widetilde{\boldsymbol{B}}_e^d \\ \overset{\circ}{\widetilde{\boldsymbol{B}}}_e = (\boldsymbol{D}^d - \boldsymbol{D}_p)\cdot\widetilde{\boldsymbol{B}}_e + \widetilde{\boldsymbol{B}}_e\cdot(\boldsymbol{D}^d - \boldsymbol{D}_p) \\ \boldsymbol{D}_p = \frac{\sigma_S}{2\eta(\tau_{eq})} \\ \eta(\tau_{eq}) = A\tau_0 \frac{(\tau_{eq}/\tau_0)}{sinh(\tau_{eq}/\tau_0)} \end{cases} \qquad [2.18]$$

where J is the volume change factor; $\mathbf{I}$ is the unity tensor; H is the strain hardening modulus; K and G are bulk and shear moduli; $\widetilde{\boldsymbol{B}}^d$ is the deviatoric part of the isochoric left Cauchy–Green deformation tensor; $\widetilde{\boldsymbol{B}}_e^d$ is the deviatoric part of the isochoric elastic left Cauchy–Green strain tensor; $\overset{\circ}{\widetilde{\boldsymbol{B}}}_e$ is the Jaumann rate of $\widetilde{\boldsymbol{B}}_e$; $\boldsymbol{D}^d$ and $\boldsymbol{D}_p$ are deviatoric and plastic parts of the rate of deformation tensor respectively; η is the viscosity that strongly depends on the equivalent shear stress τ_{eq}; and A and τ_0 are material constants.

These three models were often used to reproduce the viscoplastic behavior of polycarbonate for a range of strain rates and temperatures. Based on these models, a variant of physically-based models of polycarbonate have been developed over the recent decades (Tervoort *et al.* 1996; Engels *et al.* 2009; Varghese and Batra 2009; Van Breemen *et al.* 2011; Safari *et al.* 2012). However, the calibration of material parameters is quite difficult since large numbers of parameters are involved in the physically-based models. In addition, these models are based on an idealized microstructure, while the strain softening is still characterized in a phenomenological way.

2.1.2.2. *Phenomenological models*

Phenomenological models attempt to describe the viscoplastic behavior of polycarbonate with some considerations of the internal structure. The relationship between stress and strain is empirically modeled from the experimental observation within the framework of irreversible thermodynamics. At the phenomenological level, some classic constitutive models originally developed for metals can be modified to describe the viscoplastic behavior of polycarbonate. For example, Frank and Brockman (2001) adapted the Bodner–Partom viscoplastic model (Bodner and Partom 1975) to describe the nonlinear behavior after yielding and the strain rate-dependent behavior of polycarbonate. The Perzyna viscoplastic model (Perzyna 1966) was used by Van der Sluis *et al.* (2001) to analyze the nonlinear behavior of polycarbonate. Krempl and Ho (2000) expanded the Krempl viscoplastic model (Krempl 1995; Krempl 1996), originally developed for metals, to model the nonlinear behavior and the strain rate-dependent deformation of polymethylmethacrylate and polycarbonate. Duan *et al.* (2001) proposed a constitutive model of polycarbonate based on the Johnson–Cook model, the G'Sell–Jonas model, the Matsuoka model and the Brooks model.

When compared with the physically-based models, the constitutive equations formulated in phenomenological models are much more straightforward due to the neglecting of microstructural features. Therefore, the calibration of model parameters in phenomenological models is efficient. Many phenomenological models have been developed for polycarbonate and other polymers due to their efficiency and practicality. For detailed reviews, refer to Bouvard *et al.* (2009). However, neglecting the microstructural features makes these models fail to predict the micromechanical behavior (Anand and Ames 2006). In addition, the lack of the physical mechanism to interpret the macroscopic mechanical behavior of polycarbonate is another disadvantage of the phenomenological models.

2.2. Damage-based elastic–viscoplastic model for polycarbonate

In this section, a phenomenological constitutive model is developed (Wang *et al.* 2016) within the framework of the irreversible thermodynamics and continuum damage mechanics (CDM). An important feature of this model is that the strain softening behavior of polycarbonate is characterized by microstructural damage. Thus, this model is known as the damage-based elastic–viscoplastic model. Compared with the existing phenomenological models, the physical mechanism of the strain softening behavior is included in this model by considering the microstructural damage of polycarbonate.

2.2.1. *Damage mechanism*

Damage refers to the generation and expansion of micro-voids and micro-cracks in solid materials at the microscopic scale (Lemaitre 2012). Although damage evolution is difficult to detect directly, it can be reflected by the decrease in elastic modulus at the macroscopic scale and easily obtained using the uniaxial cyclic loading–unloading test. According to de Souza Neto *et al.* (2011), damage of amorphous polymers takes place due to straining or thermal activation. At the microscopic scale, it is characterized by the rupture of molecular bonds. In fact, the microstructure of amorphous polymers is considered as a complicated molecular network. Molecular chains are cross-linked and the branches entangle themselves randomly. To illustrate the deformation and damage mechanisms of the polymer chains, a hypothetic microstructure cell is introduced, as shown in Figure 2.10.

In this microstructure cell, two main molecular chains are cross-linked at the center point and the branches entangle each other. In the elastic regime, the molecular chains and the branches are stretched along the direction of the applied stress. When yielding starts, the molecular configuration deforms and rearranges irreversibly. Rupture of the molecular chains and disentanglement of the branches occur with increasing stress and contribute to the damage of the molecular network. The growing damage leads to the decrease in the elastic modulus and effective stress, namely the strain softening behavior of polymers. When the strain softening is completed, strain hardening behavior emerges as a result of subsequent orientation of molecular chains.

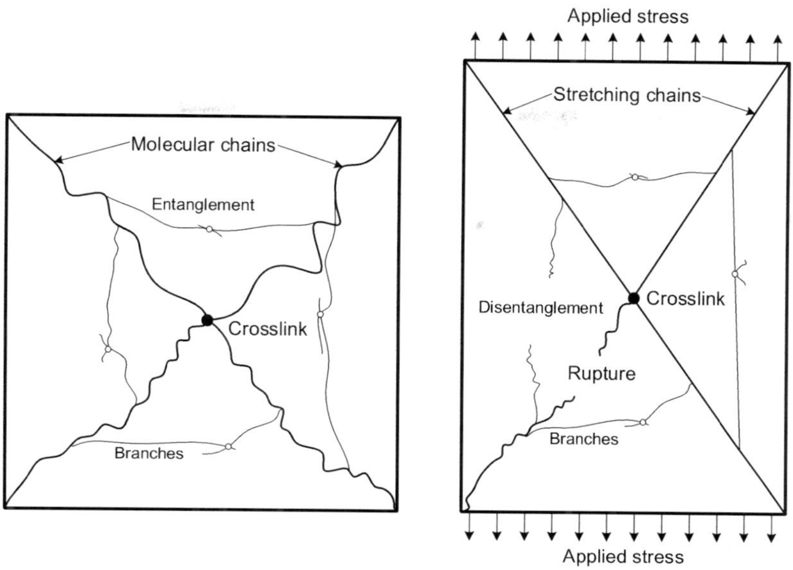

Figure 2.10. *Hypothetic microstructure cell of the amorphous polymer molecular network*

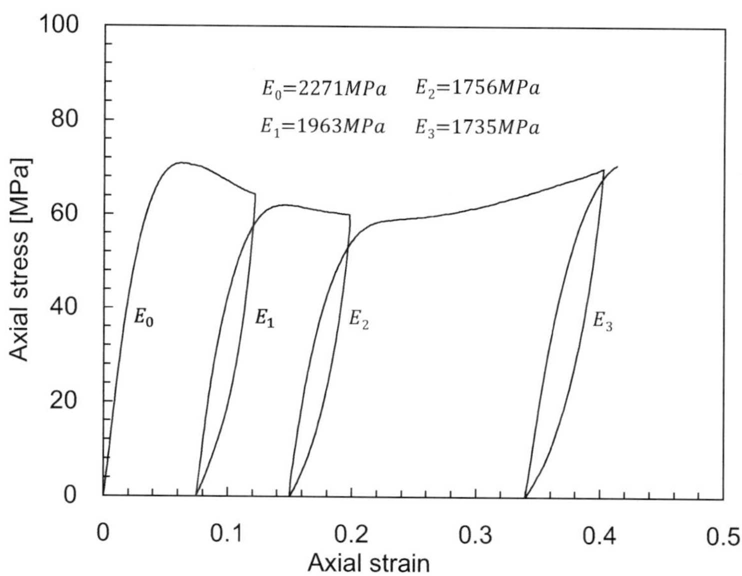

Figure 2.11. *Stress–strain curve of the polycarbonate undergoing cyclic loading and unloading*

Figure 2.11 shows the stress–strain curve of the polycarbonate obtained from the uniaxial compressive cyclic loading–unloading test. It can be seen that the elastic modulus of the polycarbonate slightly decreases after each loading–unloading cycle, indicating that internal damage of the molecular network occurs. In addition, the residual plastic strain and the viscous hysteresis loop reveal that plastic and viscous dissipations take place simultaneously. Thus, the plastic dissipation, the viscous dissipation and the damage dissipation constitute together the internal dissipation sources of the loaded polycarbonate.

A damage-based elastic–viscoplastic constitutive model is developed to describe the aforementioned characteristics of polycarbonate. The rheological representation of the model is shown in Figure 2.12. The elastic behavior is modeled by a linear Hooke's spring, while four mechanical elements, namely a dashpot, a friction device, a spring and a damage device, jointly describe the post-yield inelastic behavior of the polycarbonate. Specifically, the dashpot and the friction device describe the time-dependent viscoplasticity; the damage device and the spring describe the damage-based strain softening; the friction device and the spring describe the nonlinear strain hardening.

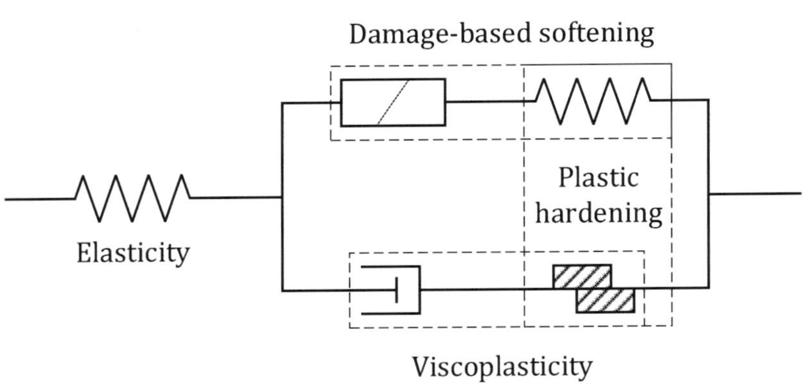

Figure 2.12. *Rheological representation of the model*

2.2.2. *Helmholtz free energy*

Within the framework of irreversible thermodynamics and CDM, constitutive equations are derived from the damage-coupled free-energy potentials (Maugin 1992). In the present study, the Helmholtz free energy is formulated in terms of the internal variables as

$$\Psi = \Psi(\boldsymbol{\varepsilon}^e, \bar{\varepsilon}^p, D, T) \qquad [2.19]$$

where $\boldsymbol{\varepsilon}^e$ is the elastic strain tensor, $\bar{\varepsilon}^p$ is the cumulated plastic strain, T is the temperature and D is the internal damage variable introduced by Kachanov (1958) to show the decrease in the elastic modulus:

$$D = 1 - \tilde{E}/E \qquad [2.20]$$

where E is the elastic modulus of the virgin material without damage and $\tilde{E}$ is the decreasing elastic modulus with growing damage. Under the hypothesis of decoupling between elasticity damage and plastic hardening (Lemaitre 2012), the Helmholtz free energy can be split into elastic and plastic parts:

$$\Psi = \Psi_e(\boldsymbol{\varepsilon}^e, D) + \Psi_p(\bar{\varepsilon}^p, T) \qquad [2.21]$$

where the first part $\Psi_e(\boldsymbol{\varepsilon}^e, D)$ is a coupled elasticity-damage strain energy, indicating that the growing damage contributes to the release of elastic strain energy. The second part $\Psi_p(\bar{\varepsilon}^p, T)$ is a plastic hardening contribution to the free energy.

2.2.3. *Fundamental laws of thermodynamics*

In continuum mechanics, a constitutive model has to satisfy the first (conservation of energy) and second (irreversibility of entropy production) laws of thermodynamics. The combination of the first and second laws of thermodynamics gives the following Clausius–Duhem form of the entropy inequality:

$$-\left(\dot{\Psi} + \eta\dot{T}\right) + \boldsymbol{\sigma}:\dot{\boldsymbol{\varepsilon}} \geq 0 \qquad [2.22]$$

where η is the entropy.

The substitution of the free-energy function [2.21] and the additive split of the strain rate $\dot{\boldsymbol{\varepsilon}} = \dot{\boldsymbol{\varepsilon}}^e + \dot{\boldsymbol{\varepsilon}}^p$ into the inequality function [2.22] gives

$$\left(\boldsymbol{\sigma} - \frac{\partial \Psi_e}{\partial \boldsymbol{\varepsilon}^e}\right) \dot{\boldsymbol{\varepsilon}}^e - \frac{\partial \Psi_e}{\partial D} \dot{D} + \boldsymbol{\sigma} : \dot{\boldsymbol{\varepsilon}}^p - \frac{\partial \Psi_p}{\partial \bar{\varepsilon}^p} \dot{\bar{\varepsilon}}^p - \left(\eta + \frac{\partial \Psi_p}{\partial T}\right) \dot{T} \geq 0 \qquad [2.23]$$

This inequality must be satisfied for any arbitrary thermodynamic process. Thus, with respect to the reversible variables such as the elastic strain tensor $\boldsymbol{\varepsilon}^e$ and the temperature T, the following constitutive laws hold (Haupt 2013):

$$\boldsymbol{\sigma} = \frac{\partial \Psi_e}{\partial \boldsymbol{\varepsilon}^e} \qquad [2.24]$$

$$\eta = -\frac{\partial \Psi_p}{\partial T} \qquad [2.25]$$

Then, the Clausius–Duhem entropy inequality can be reduced as

$$\boldsymbol{\sigma} : \dot{\boldsymbol{\varepsilon}}^p - \frac{\partial \Psi_e}{\partial D} \dot{D} - \frac{\partial \Psi_p}{\partial \bar{\varepsilon}^p} \dot{\bar{\varepsilon}}^p \geq 0 \qquad [2.26]$$

2.2.4. *Constitutive equations*

To describe the mechanical behavior of polycarbonate, the constitutive equations are supposed to contain the following characteristics:

Coupled damage elastic law

For isotropic damage in an isotropic material, the coupled damage elastic strain energy given in equation [2.21] can be written as

$$\Psi_e(\boldsymbol{\varepsilon}^e, D) = \frac{1}{2}(1 - D)\boldsymbol{\varepsilon}^e : \mathbb{C} : \boldsymbol{\varepsilon}^e \qquad [2.27]$$

where $\mathbb{C}$ is the standard isotropic elasticity tensor. The substitution of the above strain energy into equation [2.24] gives the coupled damage-elastic law

$$\boldsymbol{\sigma} = (1 - D)\mathbb{C} : \boldsymbol{\varepsilon}^e \qquad [2.28]$$

Strain rate- and temperature-dependent yield criterion

The damage-based von Mises yield criterion is used to describe the yield behavior of polycarbonate:

$$\Phi = \frac{\sqrt{3J_2(\boldsymbol{s})}}{1-D} - \sigma_{y_0} - \kappa \qquad [2.29]$$

where $\boldsymbol{s}$ is the deviatoric stress tensor, κ is the thermodynamic force associated with isotropic strain hardening and σ_{y_0} is a strain rate- and temperature-dependent initial yield stress, which is mathematically formulated as

$$\sigma_{y_0} = \sigma_{y_0}\left(\dot{\varepsilon}_{eq}, T_h\right) \qquad [2.30]$$

where $\dot{\varepsilon}_{eq} = \sqrt{(2/3)\,\dot{\boldsymbol{\epsilon}}:\dot{\boldsymbol{\epsilon}}}$ is the equivalent strain rate, $T_h = (T - T_r)/(T_g - T_r)$ is the homologous temperature, $\dot{\boldsymbol{\epsilon}}$ is the deviator of the strain increment, T is the testing temperature and T_r and T_g are the room temperature (298 K) and glass transition temperature (423 K) of polycarbonate respectively.

As indicated by Kontou (2005) and Richeton *et al.* (2006), the bilinear relationship between the initial yield stress and the logarithmic strain rate is observed in the present experiments. Here, an exponential function is introduced to approximate this relationship. In addition, the yield stress has a linear relationship with the temperature between the glass (α) transition temperature and the secondary (β) transition temperature. Accordingly, the yield stress is expressed as

$$\sigma_{y_0} = K\left(1 + C_r\dot{\varepsilon}_{eq}^{m}\right)(1 + C_tT_h) \qquad [2.31]$$

where K, C_r and m are strain rate sensitivity parameters and C_t is the temperature sensitivity parameter.

Hardening law

The thermodynamic driving force of isotropic strain hardening, κ, is defined to describe the evolution of yield stress with the cumulated plastic strain:

$$\kappa = \frac{\partial \Psi_p}{\partial \bar{\varepsilon}^p} \qquad [2.32]$$

Let us now consider the following nonlinear plastic strain hardening energy:

$$\Psi_p(\bar{\varepsilon}^p) = C_h(\bar{\varepsilon}^p)^\gamma \qquad [2.33]$$

where C_h and γ are the hardening modulus and hardening index respectively. The substitution of equation [2.33] into equation [2.32] gives rise to the thermodynamic driving force

$$\kappa = \gamma C_h(\bar{\varepsilon}^p)^{\gamma-1} \qquad [2.34]$$

Finally, the substitution of equations [2.31] and [2.34] into equation [2.29] gives the yield function

$$\Phi\left(\boldsymbol{\sigma}, D, \dot{\varepsilon}_{eq}, T, \bar{\varepsilon}^p\right) = \frac{\sqrt{3J_2(s)}}{1-D} - K\left(1 + C_r\dot{\varepsilon}_{eq}^m\right)(1 + C_tT_h) - C_h\gamma(\bar{\varepsilon}^p)^{\gamma-1} \qquad [2.35]$$

Damage evolution law

For amorphous polymers, strain softening and strain hardening are induced by the rearrangement and orientation of the molecular network at the microscopic level. This irreversible transformation of the molecular configuration could lead to micro-void nucleation, growth and micro-crack coalescence (Cayzac *et al.* 2013). Void growth and crack propagation are linked to the plastic strain and stress levels. Here, damage is characterized at the macroscopic level using the CDM approach. The experimental observations indicate that damage evolves with plastic deformation. Thus, the following explicit relation between damage and the cumulated plastic strain is adopted:

$$D = D_L\left(1 - e^{-s\bar{\varepsilon}^p}\right) \qquad [2.36]$$

where D_L is the saturated value of damage when the rearrangement of the molecular configuration is complete and s is the rate of damage evolution.

Viscoplastic flow rule

Within the framework of CDM, the flow potential of damage-coupled plasticity can be decomposed into the plastic dissipation part and the damage dissipation part:

$$\Xi = \Xi_P + \Xi_D \qquad [2.37]$$

The potential of damage dissipation Ξ_D is given by

$$\Xi_D = s(D_L - D)Y \quad [2.38]$$

where Y is the thermodynamic driving force associated with damage. It can be derived from the free energy as

$$Y = -\frac{\partial \Psi_e}{\partial D} = \frac{1}{2}\boldsymbol{\varepsilon}^e : \mathbb{C} : \boldsymbol{\varepsilon}^e \quad [2.39]$$

The plastic dissipation potential Ξ_P is assumed to be equal to the yield function Φ (Haupt 2013). Thus, the specific expression of the flow potential is obtained as

$$\Xi = \frac{\sqrt{3J_2(\boldsymbol{s})}}{1-D} - K\left(1 + C_r \dot{\varepsilon}_{eq}^m\right)(1 + C_t T_h) - C_h \gamma (\bar{\varepsilon}^p)^{\gamma-1} + s(D_L - D)Y \quad [2.40]$$

Following the generalized normality rules, the viscoplastic flow rule and the evolution laws of strain hardening and damage are given as

$$\dot{\boldsymbol{\varepsilon}}^p = \dot{\lambda}\frac{\partial \Xi}{\partial \boldsymbol{\sigma}} \quad [2.41]$$

$$\dot{\bar{\varepsilon}}^p = \dot{\lambda}\frac{\partial \Xi}{\partial \kappa} \quad [2.42]$$

$$\dot{D} = \dot{\lambda}\frac{\partial \Xi}{\partial Y} \quad [2.43]$$

The substitution of equation [2.40] into equations [2.41]–[2.43] leads to the explicit expressions

$$\dot{\boldsymbol{\varepsilon}}^p = \sqrt{\frac{3}{2}}\frac{\dot{\lambda}\boldsymbol{N}}{(1-D)} \quad [2.44]$$

$$\dot{\bar{\varepsilon}}^p = \dot{\lambda} \quad [2.45]$$

$$\dot{D} = \dot{\lambda} s(D_L - D) \quad [2.46]$$

where $\boldsymbol{N} = \boldsymbol{s}/\|\boldsymbol{s}\|$ is the flow vector of the plastic strain and $\dot{\lambda}$ is a non-negative multiplier to determine the magnitude of the viscoplastic strain, which is defined as

$$\dot{\lambda}(\boldsymbol{\sigma}, D, \sigma_y) = \begin{cases} \frac{1}{\mu}\left\{\left[\frac{\sqrt{3J_2(\boldsymbol{s})}}{(1-D)\sigma_y}\right]^{1/\xi} - 1\right\}, & \Phi(\boldsymbol{\sigma}, D, \dot{\varepsilon}_{eq}, T, \bar{\varepsilon}^p) \geq 0 \\ 0, & \Phi(\boldsymbol{\sigma}, D, \dot{\varepsilon}_{eq}, T, \bar{\varepsilon}^p) < 0 \end{cases} \quad [2.47]$$

where μ is a viscosity-related material constant and ξ is a dimensionless rate-sensitive model parameter. The yield stress $\sigma_y = \sigma_{y_0}(\dot{\varepsilon}_{eq}, T) + \kappa(\bar{\varepsilon}^p)$ is given by equations [2.31] and [2.34].

Finally, the damage-based constitutive model can be summarized as follows:

– Free-energy function:

$$\Psi = \Psi_e(\boldsymbol{\varepsilon}^e, D) + \Psi_p(\bar{\varepsilon}^p, T)$$

– Coupled damage-elastic equations:

$$\boldsymbol{\sigma} = (1 - D)\mathbb{C}: \boldsymbol{\varepsilon}^e$$

– Yield function:

$$\Phi(\boldsymbol{\sigma}, D, \dot{\varepsilon}_{eq}, T, \bar{\varepsilon}^p) = \frac{\sqrt{3J_2(\boldsymbol{s})}}{(1-D)} - K(1 + C_r\dot{\varepsilon}_{eq}^m)(1 + C_tT_h) - \gamma C_h(\bar{\varepsilon}^p)^{\gamma-1}$$

– Viscoplastic flow rule and the evolution laws of the strain hardening and damage:

$$\dot{\boldsymbol{\varepsilon}}^p = \sqrt{\frac{3}{2}}\frac{\dot{\lambda}\boldsymbol{s}}{(1-D)\|\boldsymbol{s}\|}$$

$$\dot{\bar{\varepsilon}}^p = \dot{\lambda}$$

$$\dot{D} = \dot{\lambda}s(D_L - D);$$

– Viscoplastic multiplier:

$$\dot{\lambda} = \begin{cases} \frac{1}{\mu}\left\{\left[\frac{\sqrt{3J_2(\boldsymbol{s})}}{(1-D)\sigma_y}\right]^{1/\xi} - 1\right\}, & \Phi\left(\boldsymbol{\sigma}, D, \dot{\varepsilon}_{eq}, T, \bar{\varepsilon}^p\right) \geq 0 \\ 0, & \Phi\left(\boldsymbol{\sigma}, D, \dot{\varepsilon}_{eq}, T, \bar{\varepsilon}^p\right) < 0 \end{cases}$$

2.3. Calibration of model parameters

In this section, the calibration procedure for estimating the values of the model parameters is presented. The model parameters can be classified into five categories: elasticity, yielding, hardening, damage and viscoplasticity, as listed in Table 2.1.

Elasticity	E, υ
Yielding	K, C_r, m, C_t
Hardening	γ, C_r
Damage	D_L, s
Viscoplasticity	μ, ξ

Table 2.1. *Model parameters to be calibrated*

The proposed 3D constitutive model is first simplified to its 1D version. Then, the model parameters are calibrated against the experimental data presented in Chapter 1.

– Coupled damage-elastic equations:

$$\sigma = (1 - D)E(\varepsilon - \varepsilon^p)$$

– Yield function:

$$\Phi = \frac{|\sigma|}{1-D} - K\left(1 + C_r \dot{\varepsilon}_{eq}^m\right)(1 + C_t T_h) - \gamma C_h (\bar{\varepsilon}^p)^{\gamma-1}$$

– Viscoplastic flow rule and the evolution laws of the hardening and damage:

$$\dot{\varepsilon}^p = \frac{\dot{\lambda}}{(1-D)} \operatorname{sign}(\sigma)$$

$$\dot{\bar{\varepsilon}}^p = \dot{\lambda}$$

$$\dot{D} = \dot{\lambda} s (D_L - D)$$

where the $\operatorname{sign}(\cdot)$ is a signum function to extract the sign of the Cauchy stress σ.

– Viscoplastic multiplier:

$$\dot{\lambda} = \frac{1}{\mu}\left\{\left[\frac{|\sigma|}{(1-D)\sigma_y}\right]^{1/\xi} - 1\right\}$$

The model parameters are calibrated by fitting the predicted results against the experimental data.

– Elastic modulus E is determined from the uniaxial stress–strain curve, and Poisson's ratio υ takes the value of 0.4.

– The strain rate-sensitive model parameters, K, C_r and m, are determined from the experimental data for yield stress at different strain rates, as shown in Figure 2.13.

– The temperature-sensitive model parameter C_t is determined from the experimental data for yield stress at different temperatures, as shown in Figure 2.14.

– The model parameters C_h and γ control the nonlinear strain hardening behavior of polycarbonate. Figure 2.15 shows how these two parameters affect the strain hardening behavior. The arrows in the figures indicate the increase in values.

– The saturated value of the damage D_L and the rate of the damage evolution s are determined from the experimental damage evolution curve, as shown in Figure 2.16.

– The model parameters μ and ξ control the flow stress. They are determined from the experimental flow stresses at different viscoplastic strain levels.

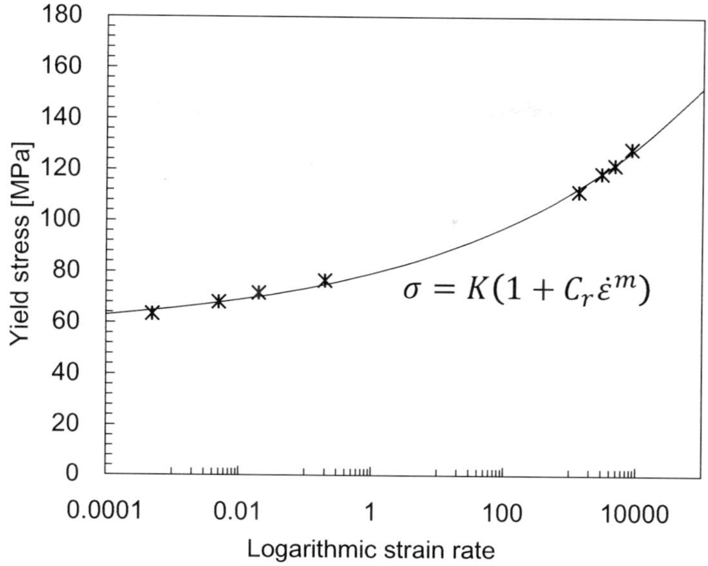

Figure 2.13. *Calibration of the model parameters K, C_r and m from the experimental data for yield stress at different strain rates*

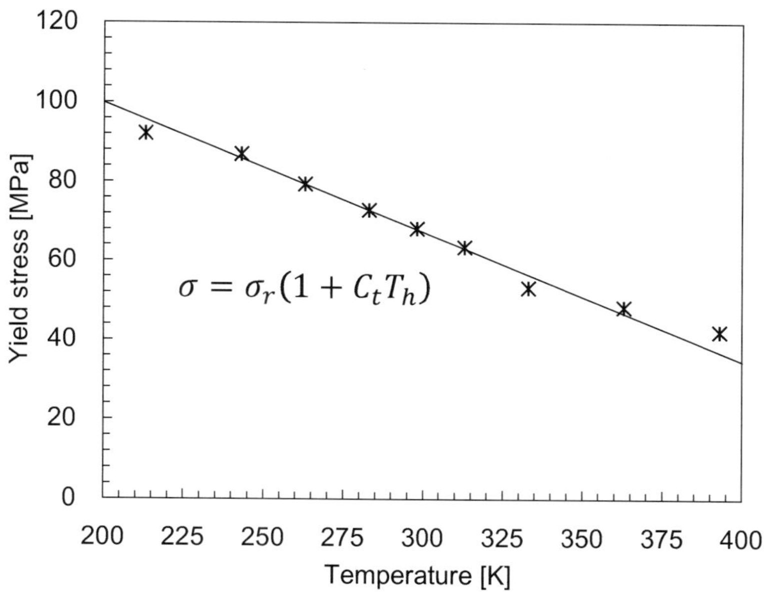

Figure 2.14. *Calibration of the model parameter C_t from the experimental data for yield stress at different temperatures*

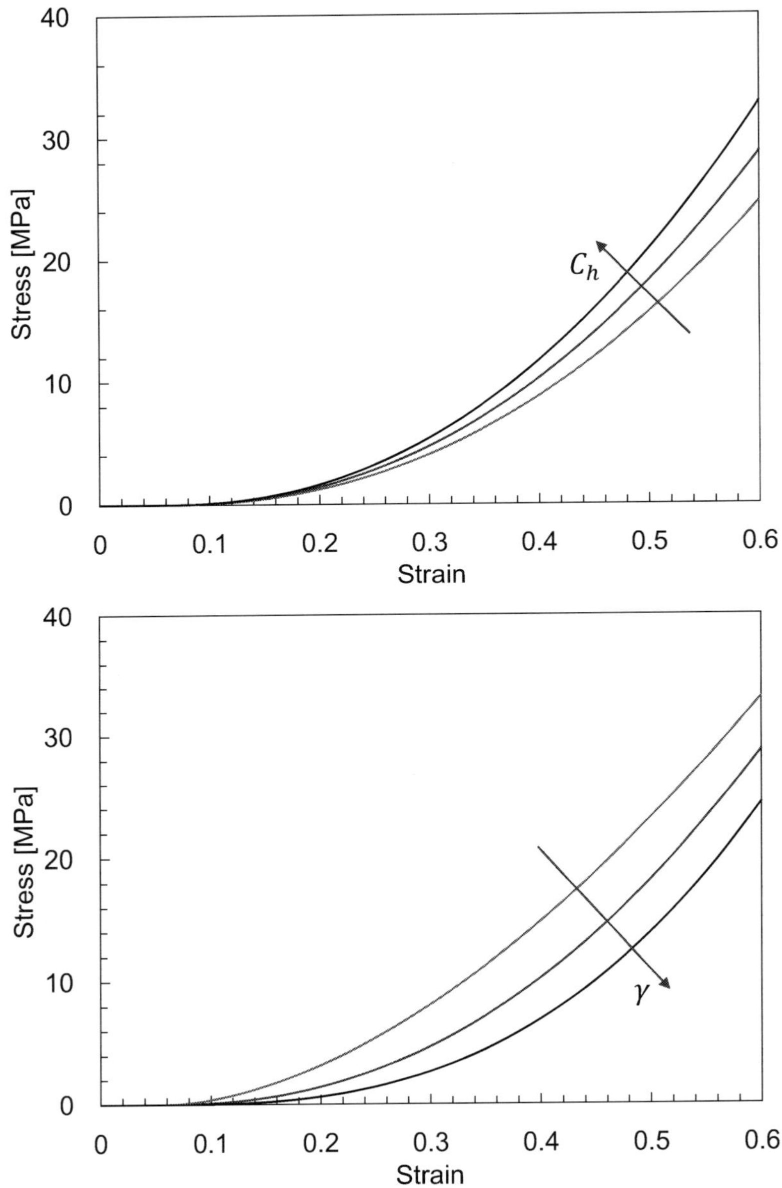

Figure 2.15. *Effects of the model parameters C_h and γ on the nonlinear strain hardening behavior; the arrows indicate the increase in values*

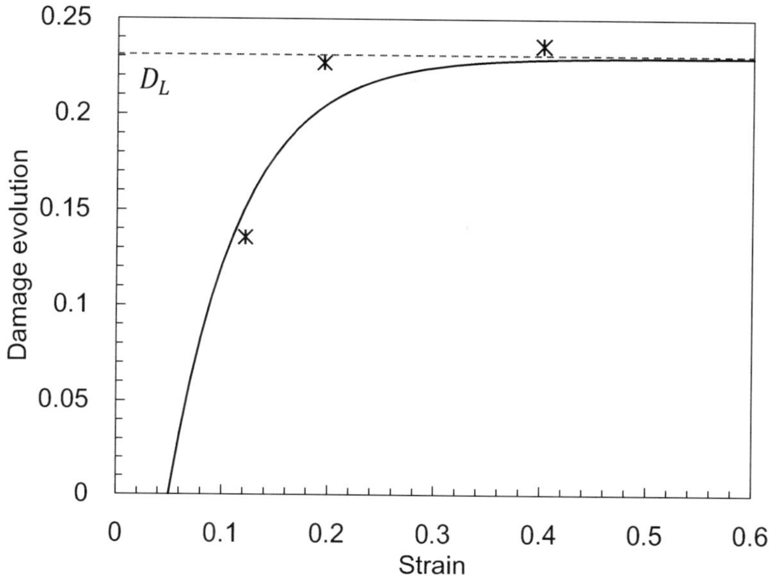

Figure 2.16. *Calibration of the model parameters D_L and s from the experimental damage evolution curve*

2.4. Numerical integration algorithm

In order to implement the constitutive model in a commercial FE code, the constitutive equations must be discretized and solved step by step through a numerical integration scheme. This section presents the numerical integration algorithm of the proposed constitutive model. First, the constitutive equations are rewritten as a time-discrete framework using the backward Euler method. Then, a coupled elastic damage predictor and viscoplastic corrector scheme is introduced. Finally, the integration scheme, namely the nonlinear equation system, is simplified into a single-scalar Newton–Raphson scheme.

2.4.1. *Time-discrete framework*

The initial value problem of the present damage-based viscoplastic constitutive model can be numerically solved as follows: given the state $(\boldsymbol{\varepsilon}_n^e, \bar{\varepsilon}_n^p, \boldsymbol{N}_n, D_n)$ at time t_n, the prescribed strain increment $\Delta\boldsymbol{\varepsilon}$, time increment Δt over the time interval $[t_n, t_{n+1}]$ and temperature T at time t_{n+1} determine the internal variables and stress field at time t_{n+1}. To this end, the constitutive equations [2.44]–[2.46] can be rewritten in the following time-discrete forms:

$$\begin{cases} \boldsymbol{\varepsilon}^e_{n+1} = \boldsymbol{\varepsilon}^e_n + \Delta\boldsymbol{\varepsilon} - \Delta\lambda\sqrt{\frac{3}{2}}\frac{\boldsymbol{N}_{n+1}}{(1-D_{n+1})} \\ \bar{\varepsilon}^p_{n+1} = \bar{\varepsilon}^p_n + \Delta\lambda \\ D_{n+1} = D_n + \Delta\lambda s(D_L - D_{n+1}) \end{cases} \qquad [2.48]$$

where the plastic flow vector $\boldsymbol{N}_{n+1}$ and the viscoplastic multiplier $\Delta\lambda$ at time t_{n+1} are given by

$$\begin{cases} \boldsymbol{N}_{n+1} = \frac{\boldsymbol{s}_{n+1}}{\|\boldsymbol{s}_{n+1}\|} \\ \Delta\lambda = \begin{cases} \frac{\Delta t}{\mu}\left\{\left[\frac{\sqrt{3J_2(\boldsymbol{s}_{n+1})}}{(1-D_{n+1})\sigma_y}\right]^{1/\xi} - 1\right\}, & \Phi \geq 0 \\ 0, & \Phi < 0 \end{cases} \\ \Phi = \frac{\sqrt{3J_2(\boldsymbol{s}_{n+1})}}{1-D_{n+1}} - K\left(1 + \frac{C_r\|\Delta\varepsilon\|}{\Delta t}\right)(1 + C_t T_h) - \gamma C_h\left(\bar{\varepsilon}^p_{n+1}\right)^{\gamma-1} \end{cases} \qquad [2.49]$$

2.4.2. *A simplified integration algorithm of the time-discrete model*

To solve the nonlinear time-discrete constitutive equations in equation [2.48], a coupled elastic-damage predictor and viscoplastic corrector scheme (de Souza Neto *et al.* 2011) is used over the time interval $[t_n, t_{n+1}]$. This scheme includes the following three parts:

Elastic trial state

Given the strain increment $\Delta\varepsilon$, the elastic trial state is formulated by freezing the dissipative internal variables:

$$(\boldsymbol{\varepsilon}^e_{n+1})^{trial} = \boldsymbol{\varepsilon}^e_n + \Delta\boldsymbol{\varepsilon} \qquad [2.50]$$

The trial stress field is then evaluated with the following hydrostatic and deviatoric parts:

$$\begin{cases} p_{n+1}^{trial} = K\left(1 - D_{n+1}^{trial}\right)(\theta_{n+1}^{e})^{trial} \\ \boldsymbol{s}_{n+1}^{trial} = 2G\left(1 - D_{n+1}^{trial}\right)(\boldsymbol{\epsilon}_{n+1}^{e})^{trial} \end{cases} \qquad [2.51]$$

where p and $\mathbf{s}$ denote the hydrostatic and deviatoric stresses respectively, K and G are bulk and shear moduli respectively, and θ and $\boldsymbol{\epsilon}$ denote the volumetric and deviatoric components of strain respectively.

Checking plastic admissibility

If the yield function satisfies $\Phi \leq 0$, the elastic trial state lies within the elastic domain, and the deformation within the time interval $[t_n, t_{n+1}]$ is purely elastic. In this case, the elastic trial state is the solution of the integration problem $(\cdot)_{n+1} = (\cdot)_{n+1}^{trial}$. Otherwise, the elastic trial state is not admissible, the viscoplastic corrector step is activated and the discrete constitutive equation [2.48] must be solved.

Simplification of the viscoplastic correcting process

In order to improve the computational efficiency, the nonlinear system expressed in equation [2.48] is simplified. As the plastic flow vector $\boldsymbol{N}$ is purely deviatoric, the combination of equations [2.48] and [2.51] gives the following stress relation between the solution state and the trial state:

$$\boldsymbol{s}_{n+1} = \left(\frac{1-D_{n+1}}{1-D_{n+1}^{trial}}\right) \boldsymbol{s}_{n+1}^{trial} - 2G\Delta\lambda\sqrt{\frac{3}{2}}\frac{\boldsymbol{s}_{n+1}}{\|\boldsymbol{s}_{n+1}\|} \qquad [2.52]$$

This means that the flow vectors of the trial state and the solution state coincide. The substitution of $\frac{\boldsymbol{s}_{n+1}}{\|\boldsymbol{s}_{n+1}\|} = \frac{\boldsymbol{s}_{n+1}^{trial}}{\|\boldsymbol{s}_{n+1}^{trial}\|}$ into equation [2.52] gives

$$\boldsymbol{s}_{n+1} = \left(\frac{1-D_{n+1}}{1-D_{n+1}^{trial}} - \sqrt{\frac{3}{2}}\frac{2G\Delta\lambda}{\|\boldsymbol{s}_{n+1}^{trial}\|}\right) \boldsymbol{s}_{n+1}^{trial} \qquad [2.53]$$

The damage equation given in equation [2.48] can then be rewritten in the following form:

$$D_{n+1}(\Delta\lambda) = \frac{D_n + \Delta\lambda s D_L}{1+\Delta\lambda s} \qquad [2.54]$$

By substituting equations [2.49] and [2.50] into the expression of the viscoplastic multiplier, the following equation can be obtained:

$$\sqrt{\frac{3}{2}}\frac{\|\boldsymbol{s}_{n+1}^{trial}\|}{(1-D_n)} - \frac{3G\Delta\lambda(1+\Delta\lambda s)}{(1-D_n)+\Delta\lambda s(1-D_L)} \\ -\left[\sigma_{y_0}(\Delta\boldsymbol{\varepsilon},\Delta t,T) + \kappa_{n+1}\left(\bar{\varepsilon}_n^p + \Delta\lambda\right)\right]\left(\frac{\Delta\lambda\mu}{\Delta t}+1\right)^{\xi} = 0 \qquad [2.55]$$

Finally, the viscoplastic correcting process can be simplified into equation [2.55], where $\Delta\lambda$ is the only unknown variable. The Newton–Raphson scheme is used to solve equation [2.55]. Once $\Delta\lambda$ is obtained, all internal variables and the stress field can be updated using the relevant equations listed above.

The overall coupled elastic-damage predictor and viscoplastic corrector scheme is summarized in Table 2.2. For finite element analysis of a boundary value problem, equilibrium equations, compatibility equations and constitutive equations must be dealt with. In this study, the equilibrium equations and the compatibility equations are handled by the commercial finite element software LS-DYNA. The constitutive equations are implemented into LS-DYNA by means of a user-defined material subroutine (UMAT).

(i) **Elastic predictor**. Given $\Delta\boldsymbol{\varepsilon}$ and state variables at t_n, evaluate the elastic trial state Elastic trial strain: $(\boldsymbol{\varepsilon}_{n+1}^e)^{trial} = \boldsymbol{\varepsilon}_n^e + \Delta\boldsymbol{\varepsilon}$ Trial cumulated plastic strain: $\left(\bar{\varepsilon}_{n+1}^p\right)^{trial} = \bar{\varepsilon}_n^p$ Trial damage: $D_{n+1}^{trial} = D_n$ Volumetric trial stress: $p_{n+1}^{trial} = K\left(1 - D_{n+1}^{trial}\right)(\theta_{n+1}^e)^{trial}$ Deviatoric trial stress: $\boldsymbol{s}_{n+1}^{trial} = 2G\left(1 - D_{n+1}^{trial}\right)(\boldsymbol{\epsilon}_{n+1}^e)^{trial}$
(ii) **Check plastic admissibility** IF $\Phi \le 0$ THEN set $(\cdot)_{n+1} = (\cdot)_{n+1}^{trial}$ and EXIT. ELSE GOTO (iii).

(iii) **Return mapping**. Solve the following nonlinear equation: $\sqrt{\frac{3}{2}}\frac{\Vert s_{n+1}^{trial}\Vert}{(1-D_n)} - \frac{3G\Delta\lambda(1+\Delta\lambda s)}{(1-D_n)+\Delta\lambda s(1-D_L)} - \left[\sigma_{y_0}(\Delta\boldsymbol{\varepsilon}, \Delta t, T) + \kappa_{n+1}(\bar{\varepsilon}_n^p + \Delta\lambda)\right]\left(\frac{\Delta\lambda\mu}{\Delta t} + 1\right)^{\xi} = 0$ The Newton–Raphson method is used to solve $\Delta\lambda$ and to correct the stress and update the state variable Stress: $\boldsymbol{\sigma}_{n+1} = \left(\frac{1-D_{n+1}}{1-D_{n+1}^{trial}} - \sqrt{\frac{3}{2}}\frac{2G\Delta\lambda}{\Vert \boldsymbol{s}_{n+1}^{trial}\Vert}\right)\boldsymbol{s}_{n+1}^{trial} + p_{n+1}^{trial}$ Damage variable: $D_{n+1} = \frac{D_n+\Delta\lambda s D_L}{1+\Delta\lambda s}$ Cumulated plastic strain: $\bar{\varepsilon}_{n+1}^p = \bar{\varepsilon}_n^p + \Delta\lambda$ Elastic strain: $\boldsymbol{\varepsilon}_{n+1}^e = \boldsymbol{\varepsilon}_n^e + \Delta\boldsymbol{\varepsilon} - \Delta\lambda\sqrt{\frac{3}{2}}\frac{\boldsymbol{N}_{n+1}}{(1-D_{n+1})}$
(iv) **EXIT**

Table 2.2. *Numerical algorithm of the overall coupled elastic-damage predictor/viscoplastic corrector scheme*

2.5. Implementation of the constitutive model in LS-DYNA

The numerical implementation of the constitutive model is realized by means of the user-defined material subroutine UMAT in LS-DYNA. To invoke UMAT, one must write a user-defined material subroutine in the UMAT interface and define a part in the keyword input deck that uses *MAT_USER_DEFINED_MATERIAL_MODELS with appropriate input parameters. Up to 10 user subroutines can be concurrently implemented to update the stresses in solids, shells, beams and truss beams. All the subroutines are collected in the file dyn21.F.

The subroutine interface is given below:

```
subroutine umat41 (cm,eps,sig,epsp,hsv,dt1,capa,etype,tt,
   1 temper,failel,crv)

c main context

      return
      end
```

The definition of the variables used in this subroutine is as follows:

– cm(1)-cm(n): first to nth material constants;

– eps(1)-eps(6): six components of the strain increments;

– sig(1)-sig(6): six components of the stresses in the previous time step;

– epsp: effective plastic strain in the previous time increment;

– hsv(1)-hsv(n): first to nth history variables in the previous time increment;

– dt1: current time step size;

– capa: reduction factor for transverse shear;

– etype: character string that specifies the element type;

– tt: current time;

– temper: current temperature;

– failel: flag indicating failure of the element;

– crv(101,2,*): array representation of curves defined in the keyword deck.

The UMAT subroutine of the proposed damage-based elastic–viscoplastic constitutive model is given for the numerical algorithm presented in Table 2.2.

```
      subroutine umat41 (cm,eps,sig,epsp,hsv,dt1,capa,etype,tt,
     1 temper,failel,crv,cma)

      character*(*) etype
      dimension  cm(*),eps(*),sig(*),hsv(*),crv(101,2,*)
      logical failel

c-------------------------model parameters--------------------------
c     cm(1) = elastic modulus
c     cm(2) = Poisson's ratio
c     cm(3) = first strain rate sensitive parameter
c     cm(4) = second strain rate sensitive parameter
c     cm(5) = third strain rate sensitive parameter
c     cm(6) = temperature sensitive parameter
c     cm(7) = first hardening parameter
c     cm(8) = second hardening parameter
c     cm(9) = saturated value of damage
c     cm(10) = rate of damage growth
c     cm(11) = first viscoplastic flow parameter
c     cm(12) = second viscoplastic flow parameter
c     cm(13) = reference temperature
c     cm(14) = glass transition temperature

c------------------------history variables--------------------------
c     hsv(1)-hsv(6) = six components of plastic strain
c     hsv(7) = damage variable
c     hsv(8) = equivalent strain rate

c--------------------------main program----------------------------

C     define constants
      parameter(zero=0.,one=1.,two=2.,three=3.,six=6.,half=0.5,
     1  onehalf=1.5,tiny=1.e-10)

c     compute lame constants
      xmu    = cm(1)/(one+cm(2))/two
      xk     = cm(1)/(one-two*cm(2))/three
```

```
c    history variables
     peps1= hsv(1)
     peps2= hsv(2)
     peps3= hsv(3)
     peps4= hsv(4)
     peps5= hsv(5)
     peps6= hsv(6)
     dmg  = hsv(7)

c    compute volumetric and deviatoric strain increments
     epsincv = (eps(1)+eps(2)+eps(3))/three
     epsincd1= eps(1)-epsincv
     epsincd2= eps(2)-epsincv
     epsincd3= eps(3)-epsincv
     epsincd4= eps(4)
     epsincd5= eps(5)
     epsincd6= eps(6)

c    compute equivalent strain rate
     eepsinc= sqrt(two/three*(epsincd1**2+epsincd2**2+epsincd3**2
     1         +two*(epsincd4**2+epsincd5**2+epsincd**2)))
     esr= eepsinc/dt1

c    compute homologous temperature
     th= (temper-cm(13))/(cm(14)-cm(13))

c    compute volumetric and deviatoric stresses
     sigv = (sig(1)+sig(2)+sig(3))/three
     sigd1= sig(1)-sigv
     sigd2= sig(2)-sigv
     sigd3= sig(3)-sigv
     sigd4= sig(4)
     sigd5= sig(5)
     sigd6= sig(6)

c    compute trial stress increments
```

```
      tsigincv = three*xk*epsincv
      tsigincd1= two*xmu*epsincd1
      tsigincd2= two*xmu*epsincd2
      tsigincd3= two*xmu*epsincd3
      tsigincd4= two*xmu*epsincd4
      tsigincd5= two*xmu*epsincd5
      tsigincd6= two*xmu*epsincd6

c     compute trail stresses
      tsigv = sigv+tsigincv
      tsigd1= sigd1+tsigincd1
      tsigd2= sigd2+tsigincd2
      tsigd3= sigd3+tsigincd3
      tsigd4= sigd4+tsigincd4
      tsigd5= sigd5+tsigincd5
      tsigd6= sigd6+tsigincd6

c     compute flow vector
      tsignorm=sqrt(tsigd1**2+tsigd2**2+tsigd3**2
     1         +two*(tsigd4**2+tsigd5**2+tsigd**2))

      flowvec1= tsigd1/(tsignorm+tiny)
      flowvec2= tsigd2/(tsignorm+tiny)
      flowvec3= tsigd3/(tsignorm+tiny)
      flowvec4= tsigd4/(tsignorm+tiny)
      flowvec5= tsigd5/(tsignorm+tiny)
      flowvec6= tsigd6/(tsignorm+tiny)

c     compute von mises stress
      tsigvm=sqrt(onehalf*(tsigd1**2+tsigd2**2+tsigd3**2
     1         +two*(tsigd4**2+tsigd5**2+tsigd**2)))

c     compute yield stress
      ysig1= cm(3)*(1+cm(4)*esr**cm(5))*(1+cm(6)*th)
      ysig2= cm(7)*cm(8)*epsp**(cm(8)-1)
      ysig= ysig1+ysig2
```

```
c    yield function
     yf= tsigvm/(one-dmg)-ysig

c    check yielding and compute viscoplastic multiplier
     if(yf .le. zero) then
              pmtp= zero
     else
              call cptpmtp(tsigvm,dmg,cm(9),cm(10),cm(11),cm(12),ysig,pmtp)
     end if

c    update effective plastic strain
     epsp= epsp+pmtp

c    update damage variable
     dmg= (dmg+pmtp*cm(10)*cm(9))/(one+pmtp*cm(10))

c    compute plastic strains
     peps1=peps1+pmtp*sqrt(onehalf)*flowvec1
     peps2=peps2+pmtp*sqrt(onehalf)*flowvec2
     peps3=peps3+pmtp*sqrt(onehalf)*flowvec3
     peps4=peps4+pmtp*sqrt(onehalf)*flowvec4
     peps5=peps5+pmtp*sqrt(onehalf)*flowvec5
     peps6=peps6+pmtp*sqrt(onehalf)*flowvec6

c    compute true stresses
     sigd1=tsigd1-two*xmu*pmtp*sqrt(onehalf)*flowvec1
     sigd2=tsigd2-two*xmu*pmtp*sqrt(onehalf)*flowvec2
     sigd3=tsigd3-two*xmu*pmtp*sqrt(onehalf)*flowvec3
     sigd4=tsigd4-two*xmu*pmtp*sqrt(onehalf)*flowvec4
     sigd5=tsigd5-two*xmu*pmtp*sqrt(onehalf)*flowvec5
     sigd6=tsigd6-two*xmu*pmtp*sqrt(onehalf)*flowvec6

c    update stress vector
     sig(1)= sigd1+tsigv
     sig(2)= sigd2+tsigv
     sig(3)= sigd3+tsigv
     sig(4)= sigd4
```

```
      sig(5)= sigd5
      sig(6)= sigd6

c     update history variables
      hsv(1)= peps1
      hsv(2)= peps2
      hsv(3)= peps3
      hsv(4)= peps4
      hsv(5)= peps5
      hsv(6)= peps6
      hsv(7)= dmg
      hsv(8)= esr

c-----------------------------done--------------------------------

      return
      end
```

2.6. Numerical examples

In this section, some numerical examples are presented to demonstrate the capabilities of the damage-based elastic–viscoplastic model. First, a single-element simulation is carried out to validate the numerical efficiency and practicability of the proposed algorithmic scheme. Model predictions are compared to the experimental data. Then, the uniaxial compressive tests of the cylinder specimen are simulated.

The software LS-DYNA is used in the finite element simulation. The damage-based elastic–viscoplastic model is implemented into LS-DYNA.

Material properties, such as elastic modulus E, Poisson's ratio ν, density ρ and glass transition temperature T_g, as well as model parameters are calibrated from the experimental data and listed in Table 2.3.

Parameter	Value/unit	Parameter	Value/unit
E	*2271 MPa*	ν	*0.4*
ρ	*1190 kg/m*^{3}	T_g	*423 K*
K	*55.18 MPa*	C_r	*0.438 s*
C_t	*−0.527*	m	*0.121*
C_h	*35.29 MPa*	γ	*3.236*
μ	*0.02 s*	ξ	*1.0*
s	*15.32*	D_L	*0.229*

Table 2.3. *Material properties and model parameters*

2.6.1. *Single-element tests*

A single hexahedral element, namely eight-node reduced integration element SOLID 164 with a dimension of 1 mm × 1 mm × 1 mm, is considered and the damage-based elastic–viscoplastic model is used in the simulations. Load and boundary conditions are shown in Figure 2.17. In order to simulate a pure uniaxial compressive loading, displacement is applied on the top surface of the element. Out-plane movements on the bottom surface XOY and the side surfaces XOZ and YOZ are fully restricted.

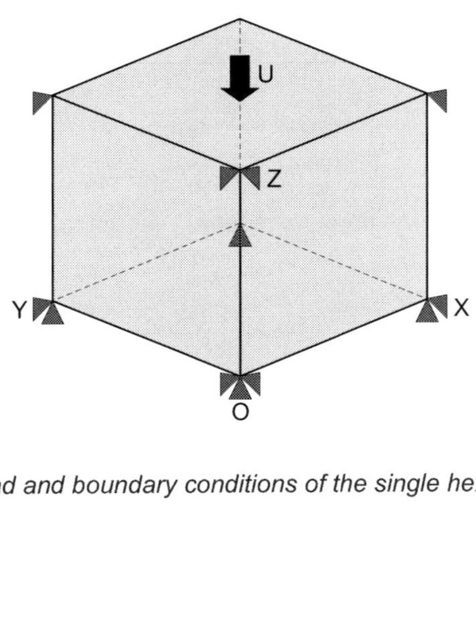

Figure 2.17. *Load and boundary conditions of the single hexahedral element*

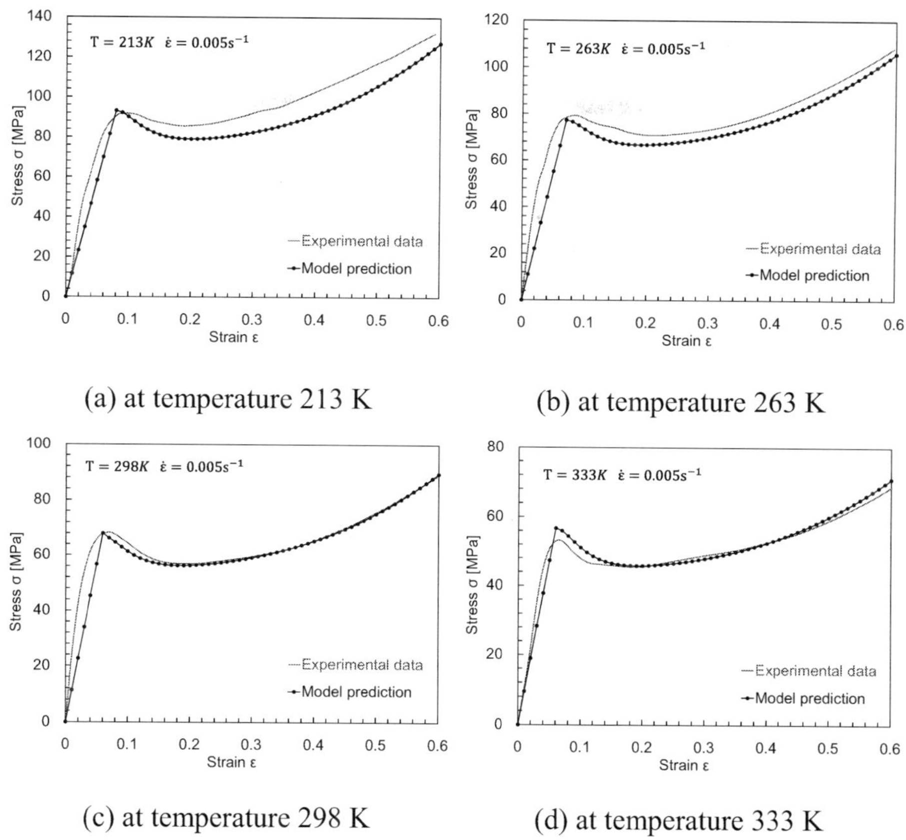

Figure 2.18. *Comparison of the predicted stress–strain responses to the experimental data at different temperatures. For a color version of this figure, see www.iste.co.uk/zhang/polycarbonate.zip*

Figure 2.18 compares the model predictions to the experimental data at the strain rate of 0.005 s^{-1} and four different temperatures of 213, 263, 298 and 333 K. It can be observed that yield stress decreases with the increase in temperature. Correspondingly, the post-yield viscoplastic flow stress also shows a downward trend. At all temperatures, the model predictions agree well with the experimental data, indicating that the mechanical behavior of polycarbonate is well described by the proposed constitutive model, especially the temperature dependence.

Figure 2.19 compares the model predictions to the experimental data at the temperature of 298 K and the strain rates of 0.005 and 0.2 s^{-1}. When the strain rate is increased from 0.005 to 0.2 s^{-1}, yield stress increases correspondingly from 68 to 75 MPa. The post-yield viscoplastic flow stress also shifts up. Overall, the strain rate-dependent material behavior of polycarbonate is reasonably captured by the proposed model. Here it should be noted that the stress is calculated by dividing the summed reaction force by the top surface area.

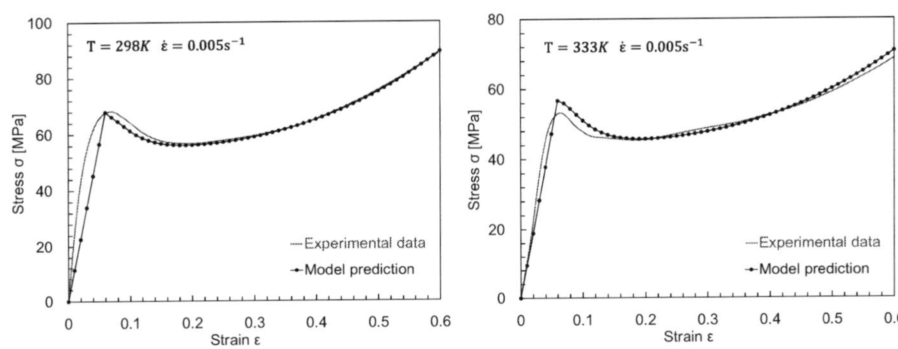

(a) at strain rate 0.005 s^{-1} (b) at strain rate 0.2 s^{-1}

Figure 2.19. *Comparisons of the predicted stress–strain responses to the experimental data at different strain rates. For a color version of this figure, see www.iste.co.uk/zhang/polycarbonate.zip*

Figure 2.20 shows the yield stress and post-yield stress drop of polycarbonate. Figure 2.20(a) compares the model predictions to the experimental data over a wide range of strain rates. Figure 2.20(b) compares the model predictions to the experimental data over the testing temperature range. The model predictions agree well with the experimental data by capturing the strain rate- and temperature-dependent yield stress, as well as the damage-based strain softening behavior. However, at a high stress level, the predicted stress drop tends to become larger than the experimental one. This is mainly because the stress drop in the present model is assumed to be related to the stress level and damage. Thus, when material damage reaches its saturated value D_L, the stress drop solely depends on the stress level and therefore cannot keep constant, as shown in the experiments.

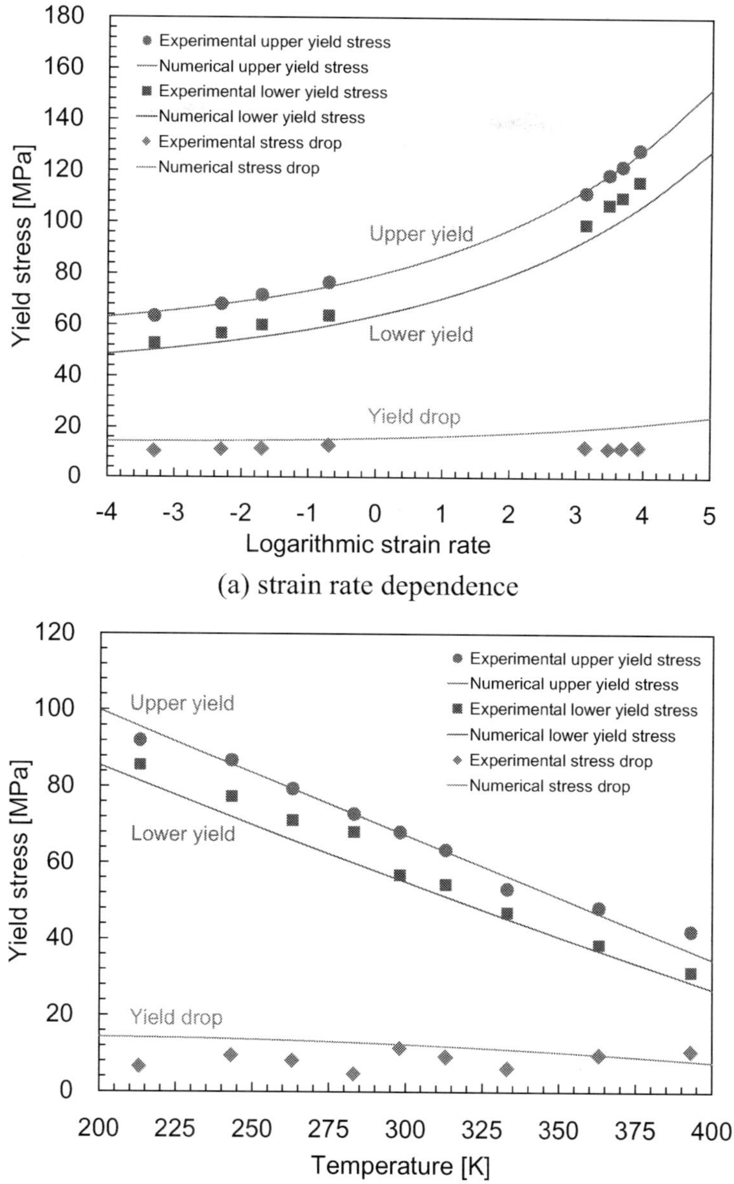

(a) strain rate dependence

(b) temperature dependence

Figure 2.20. *Comparisons of the predicted yield stress and post-yield stress drop to the experimental data. For a color version of this figure, see www.iste.co.uk/zhang/polycarbonate.zip*

Figure 2.21 shows the predicted yield stress at different strain rates and temperatures. The first observation is that the yield stress increases with increasing strain rates over the temperature range. At a low temperature (213 K), the yield stress dramatically increases with the strain rate. However, when the temperature is increased to 393 K, the increase in yield stress with the strain rate becomes weak. These model predictions are in accordance with the thermal-activated β transition behavior as well as the experimental data. In general, high strain rate and low temperature lead to a high yield stress of polycarbonate and vice versa. The present model reasonably captures the strain rate- and temperature-dependent yield behavior.

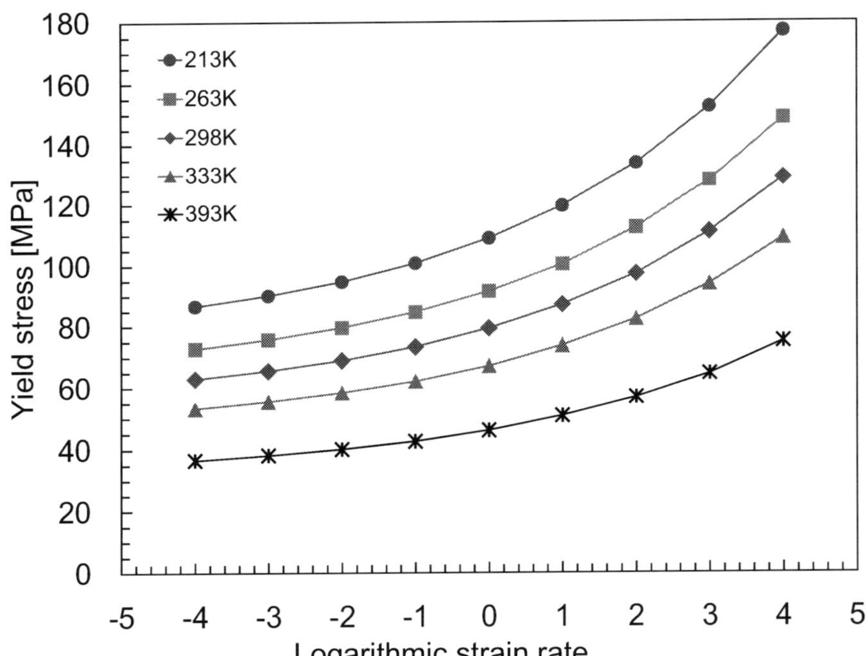

Figure 2.21. *Predicted yield stress as a function of strain rate at different temperatures. For a color version of this figure, see www.iste.co.uk/zhang/polycarbonate.zip*

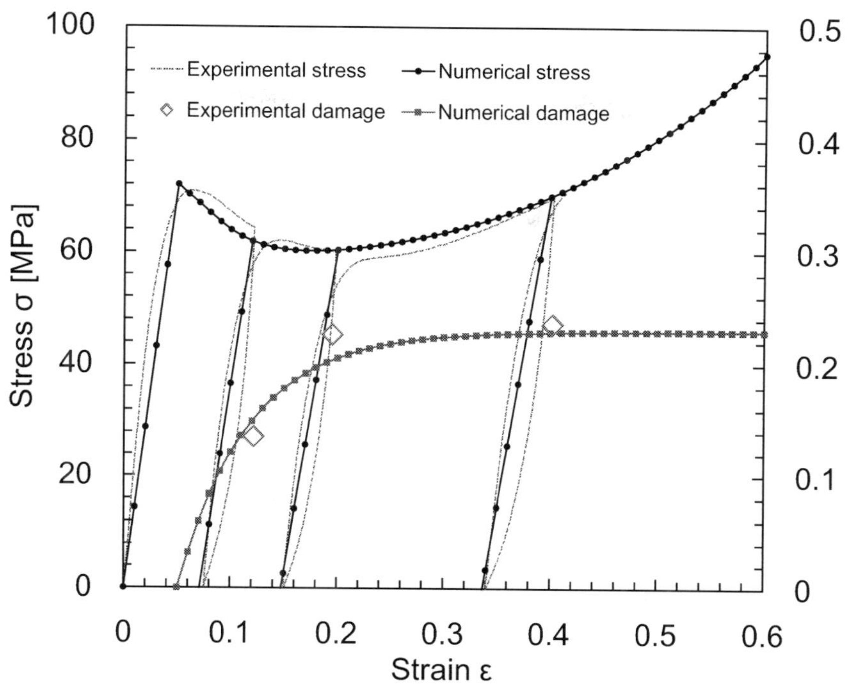

Figure 2.22. *Cyclic stress–strain response and the damage evolution at the strain rate of 0.005 s^{-1} and the temperature of 298 K. For a color version of this figure, see www.iste.co.uk/zhang/polycarbonate.zip*

The predicted cyclic stress–strain response and the damage evolution are compared to the experimental data at a strain rate of 0.005 s^{-1} and a temperature of 298 K in Figure 2.22. It can be observed that the present model depicts well the cyclic loading–unloading response and damage evolution of polycarbonate. Material damage grows with the accumulated viscoplastic strain, contributing to the post-yield stress drop. When damage reaches its saturated value D_L, the strain softening behavior completes and the subsequent strain hardening behavior becomes dominant. At the microscopic scale, the molecular defects complete rearrangement and reach a relatively stable configuration at this moment, while the orientation of molecular chains emerges.

2.6.2. *Cylinder compression simulation*

In this section, a uniaxial compressive test on the cylinder specimen is simulated at the strain rate of 0.2 s^{-1} and the temperature of 298 K. Figure 2.23 shows the mesh and boundary conditions of the compression simulation. The cylinder specimen is meshed using the eight-node reduced integration elements (LS-DYNA SOLID 164), and the number of elements is 3,000. The material model of the cylinder specimen is the damage-based elastic–viscoplastic model. Let us assume that the loaded and fixed plates are rigid bodies. The bottom plate is fixed in all directions and rotations. A time-dependent displacement is applied on the top plate. In order to reduce the computation burden, only a quarter of the cylinder specimen is simulated and symmetric boundary conditions are applied on the cutting plane.

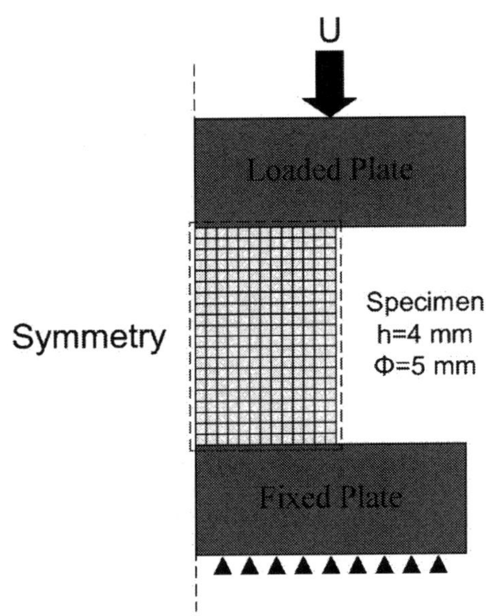

Figure 2.23. *Mesh and boundary conditions of the compression simulations*

Figure 2.24 shows the damage and the von Mises stress contours on the cylinder specimen. The cylinder specimen is compressed into a drum in the simulation, which is coincident with the experiment. Damage and von Mises stress contours are shown in Figure 2.24(a) and (b) respectively.

As noted from the figures, the material damage reaches its saturated value D_L at the center of the specimen. Similarly, the maximum von Mises stress emerges at the center of the specimen. The cross shape in the specimen looks like the plastic "slip band", indicating that the damage evolution is jointly determined by the stress level and the accumulated viscoplastic strain.

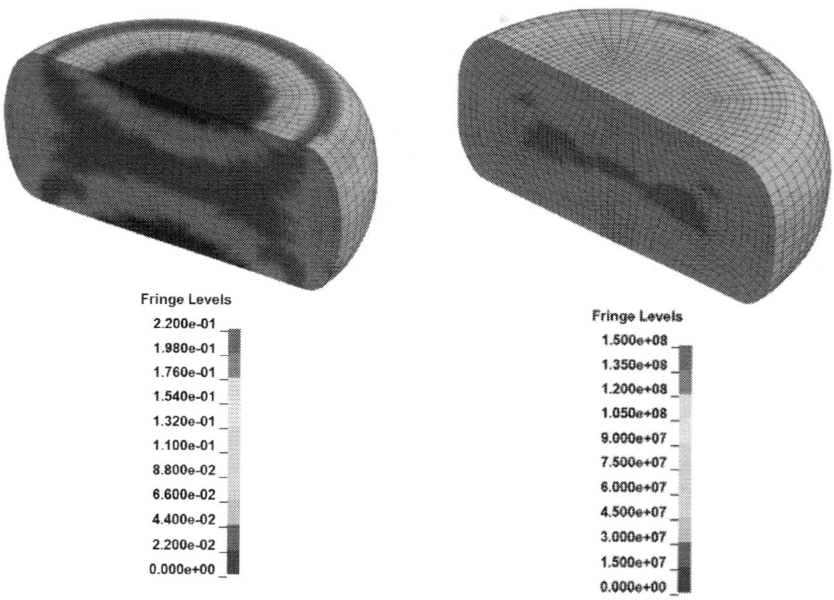

(a) damage (b) von Mises stress

Figure 2.24. *Predicted damage and von Mises stress contours on the cylinder specimen. For a color version of this figure, see www.iste.co.uk/zhang/polycarbonate.zip*

2.7. Conclusion

In this chapter, constitutive models of polycarbonate were presented. A damage-based elastic–viscoplastic constitutive model was developed for polycarbonate within the framework of irreversible thermodynamics and CDM. It was able to characterize the mechanical behavior and damage evolution of the material over a wide range of strain rates and temperatures. The constitutive equations, specifically the viscoplastic flow rule and damage evolution law, were derived from the Helmholtz free energy and the damage-coupled plastic flow potential. To realize the numerical implementation of the proposed model, a coupled elastic-damage predictor and viscoplastic corrector algorithmic scheme was introduced and further simplified into a single-scalar Newton–Raphson scheme. The model was implemented into LS-DYNA by using the UMAT subroutine. Finally, numerical examples, including both the single-element and the cylinder specimen, were given. The good correlation between the numerical predictions and the experimental data has demonstrated the capabilities of the proposed model in capturing the mechanical behavior of polycarbonate over a wide range of strain rates and temperatures.

2.8. References

Anand, L. and Ames, N.M. (2006). On modeling the micro-indentation response of an amorphous polymer. *International Journal of Plasticity*, 22(6), 1123–1170.

Argon, A.S. (1973). A theory for the low-temperature plastic deformation of glassy polymers. *Philosophical Magazine*, 28(4), 839–865.

Arruda, E.M. and Boyce, M.C. (1993). Evolution of plastic anisotropy in amorphous polymers during finite straining. *International Journal of Plasticity*, 9(6), 697–720.

Arruda, E.M., Boyce, M.C. and Jayachandran, R. (1995). Effects of strain rate, temperature and thermomechanical coupling on the finite strain deformation of glassy polymers. *Mechanics of Materials*, 19(2–3), 193–212.

Bodner, S. and Partom, Y. (1975). Constitutive equations for elastic-viscoplastic strain-hardening materials. *Journal of Applied Mechanics*, 42(2), 385–389.

Bouvard, J.L., Ward, D.K., Hossain, D., Nouranian, S., Marin, E.B. and Horstemeyer, M.F. (2009). Review of hierarchical multiscale modeling to describe the mechanical behavior of amorphous polymers. *Journal of Engineering Materials and Technology*, 131(4), 041206.

Boyce, M.C., Parks, D.M. and Argon, A.S. (1988). Large inelastic deformation of glassy polymers. Part I: rate dependent constitutive model. *Mechanics of Materials*, 7(1), 15–33.

Buckley, C.P. and Jones, D.C. (1995). Glass-rubber constitutive model for amorphous polymers near the glass transition. *Polymer*, 36(17), 3301–3312.

Cayzac, H.A., Saï, K. and Laiarinandrasana, L. (2013). Damage based constitutive relationships in semi-crystalline polymer by using multi-mechanisms model. *International Journal of Plasticity*, 51, 47–64.

Christensen, R. (2012). *Theory of Viscoelasticity: An Introduction*. Elsevier, Amsterdam.

De Focatiis, D., Embery, J. and Buckley, C.P. (2010). Large deformations in oriented polymer glasses: experimental study and a new glass-melt constitutive model. *Journal of Polymer Science Part B*, 48(13), 1449–1463.

Duan, Y., Saigal, A., Greif, R. and Zimmerman, M.A. (2001). A uniform phenomenological constitutive model for glassy and semicrystalline polymers. *Polymer Engineering & Science*, 41(8), 1322–1328.

Engels, T.A.P., Govaert, L.E. and Meijer, H.E.H. (2009). The influence of molecular orientation on the yield and post-yield response of injection-molded polycarbonate. *Macromolecular Materials and Engineering*, 294, 821–828.

Flügge, W. (2013). *Viscoelasticity*. Springer Science & Business Media, Berlin.

Frank, G.J. and Brockman, R.A. (2001). A viscoelastic-viscoplastic constitutive model for glassy polymers. *International Journal of Solids and Structures*, 38, 5149–5164.

Govaert, L.E., Timmermans, P. and Brekelmans, W. (2000). The influence of intrinsic strain softening on strain localization in polycarbonate: modeling and experimental validation. *Journal of Engineering Materials and Technology*, 122(2), 177–185.

Haupt, P. (2013). *Continuum Mechanics and Theory of Materials*. Springer Science & Business Media, Berlin.

Kachanov, L.M. (1958). Time of the rupture process under creep conditions, *Izy Akad. Nank SSR Otd Tech Nauk*, 8, 26–31.

Klompen, E.T.J., Engels, T.A.P. and Govaert, L.E. (2005). Modeling of the postyield response of glassy polymers: influence of thermomechanical history. *Macromolecules*, 38(16), 6997–7008.

Kontou, E. (2005). Effect of thermal treatments on the yielding of polycarbonate. *Journal of Applied Polymer Science*, 98(2), 796–805.

Krempl, E. (1995). The overstress dependence of inelastic rate of deformation inferred from transient tests. *Journal of the Society of Materials Science*, 44, 3–10.

Krempl, E. (1996). A small strain viscoplasticity theory based on overstress. *Unified Constitutive Laws of Plastic Deformation*, 31, 281–318.

Krempl, E. and Ho, K. (2000). An overstress model for solid polymer deformation behavior applied to Nylon 66. *ASTM STP 1357*, 118–137.

Lemaitre, J. (2012). *A Course on Damage Mechanics*. Springer Science & Business Media, Berlin.

Maugin, G.A. (1992). *The Thermomechanics of Plasticity and Fracture*. Cambridge University Press, Cambridge.

Perzyna, P. (1966). Fundamental problems in viscoplasticity. *Advances in Applied Mechanics*, 9, 243–377.

Richeton, J., Ahzi, S., Vecchio, K.S., Jiang, F.C. and Adharapurapu, R.R. (2006). Influence of temperature and strain rate on the mechanical behavior of three amorphous polymers: characterization and modeling of the compressive yield stress. *International Journal of Solids and Structures*, 43(7–8), 2318–2335.

Safari, K.H., Zamani, J., Ferreira, F.J. and Guedes, R.M. (2012). Constitutive modeling of polycarbonate during high strain rate deformation. *Polymer Engineering & Science*, 53, 752–761.

Souza Neto, E.A.de, Peric, D. and Owen, D.R.J. (2011). *Computational Methods for Plasticity: Theory and Applications*. John Wiley & Sons, New York.

Tervoort, T.A., Klompen, E.T.J. and Govaert, L.E. (1996). A multi-mode approach to finite, three-dimensional, nonlinear viscoelastic behavior of polymer glasses. *Journal of Rheology*, 40(5), 779–797.

Tervoort, T.A., Smit, R.J.M., Brekelmans, W.A.M. and Govaert, L.E. (1997). A constitutive equation for the elasto-viscoplastic deformation of glassy polymers. *Mechanics of Time-Dependent Materials*, 1(3), 269–291.

Van Breemen, L.C.A., Klompen, E.T.J., Govaert, L.E. and Meijer, H.E.H. (2011). Extending the EGP constitutive model for polymer glasses to multiple relaxation times. *Journal of the Mechanics and Physics of Solids*, 59, 2191–2207.

Van der Sluis, O., Schreurs, P.J. and Meijer, H.E. (2001). Homogenisation of structured elastoviscoplastic solids at finite strains. *Mechanics of Materials*, 33(9), 499–522.

Varghese, A.G. and Batra, R.C. (2009). Constitutive equations for thermomechanical deformations of glassy polymers. *International Journal of Solids and Structures*, 46, 4079–4094.

Wang, M.C. and Guth, E. (1952). Statistical theory of networks of non-Gaussian flexible chains. *The Journal of Chemical Physics*, 20(7), 1144–1157.

Wang, J., Xu, Y.J., Zhang, W.H. and Moumni, Z. (2016). A damage-based elastic-viscoplastic constitutive model for amorphous glassy polycarbonate polymers. *Materials & Design*, 97, 519–531.

Wu, J.J. and Buckley, C.P. (2004). Plastic deformation of glassy polystyrene: a unified model of yield and the role of chain length. *Journal of Polymer Science Part B*, 42(11), 2027–2040.

3

Impact Simulation of Polycarbonates in Aeronautical and Aerospace Applications

The impact of polycarbonate products is a subject of critical interest in aviation and aerospace applications since high-velocity impact threats are unavoidable. For example, bird strikes often result in cracking or perforation of the windshield and canopy. For astronauts, the impact of space debris is an extremely severe threat since the impact energy of the debris with only 1 mm size could reach 71 J in space. Therefore, it is of great importance to understand and to accurately predict the impact responses and catastrophic failure of polycarbonate products.

The impact behavior of polycarbonate has been studied experimentally, ranging from high-velocity impacts (Wright *et al.* 1993; Ball and McKenzie 1994; Li and Goldsmith 1997; Qasim and Yousif 2008; Gunnarsson *et al.* 2009; Qasim 2009) using gas gun impact tests to low-velocity impacts (Fraser and Ward 1977; Cheng *et al.* 1994; Lombardo *et al.* 1994; Inberg and Gaymens 2002) using falling weight tests and Izod impact tests. However, considering the complex geometries of products and the extreme service condition, e.g. the large range of strain rates from 10^{-4} to 10^{4} s^{-1} or even higher in aeronautical and aerospace applications, the experiment is extremely time-consuming.

The finite element (FE) method provides an alternative in product analysis and design, e.g. dynamic analyses of polycarbonate under impact loading (Tvergaard and Needleman 2008; Dorogoy *et al.* 2011; Hu *et al.* 2013; Dorogoy and Rittel 2015). However, the impact simulation of aeronautical and aerospace polycarbonate products remains challenging due to the complex material properties and service conditions.

The purpose of this chapter is to develop an FE simulation approach aiming to predict the high-velocity impact behavior of polycarbonate products. Within the commercial software LS-DYNA, the elastic–viscoplastic constitutive model proposed in Chapter 2 is implemented together with a failure criterion to simulate the dynamic response and failure of impacted polycarbonate products.

First, the validity of the impact simulation is verified by comparing with the gas gun impact test results of polycarbonate plates. Then, FE simulations are performed to analyze the dynamic responses of aircraft polycarbonate windshields against bird strikes and astronaut helmet visors under debris impact. The critical impact energy, maximum deformation and stress level are determined correspondingly.

3.1. Simulation methodology and experimental verification

3.1.1. *Impact simulation methodology*

Impact is a typical nonlinear problem, which can be solved by implicit and explicit methods. The nonlinear dynamic responses are incrementally obtained in both methods. However, in each increment, the solution procedures of the two methods are quite different.

The implicit method uses a Newton iterative approach to update the displacements and velocities in each increment (Crisfield *et al.* 2012). Thus, the equilibrium equations are iteratively solved in each increment until convergence is achieved. Although the iterative solution procedure ensures that a large time increment is used, the computational cost is extremely high.

The explicit method uses a central difference scheme to update the displacements and velocities (Crisfield *et al.* 2012). Equilibrium equations can be directly solved in each increment without iteration. Thus, the computational cost is relatively low, especially for the impact problem (Sun *et al.* 2000; Harewood and McHugh 2007; Crisfield *et al.* 2012). However, the size of the time increment should be small enough to stabilize the solution procedure. A conservative estimation of the stable time increment is given in (Sun *et al.* 2000; Harewood and McHugh 2007):

$$\Delta t \leq \min\left(\frac{L_e}{c_d}\right) \qquad [3.1]$$

where Δt is the time increment, L_e is the characteristic element length and c_d is the dilatational wave speed of the material (Harewood and McHugh 2007):

$$c_d = \sqrt{\frac{\lambda + 2\mu}{\rho}} \qquad [3.2]$$

where λ and μ are the Lame elastic constants and ρ is the material density.

Table 3.1 provides a comparison between the two methods. For more details, refer to (Crisfield *et al.* 2012).

Implicit method	Explicit method
Solve the equilibrium equations iteratively in each increment	Solve the equilibrium equations directly in each increment without iteration
Time increment can be large	Time increment should be smaller than a critical value
Computational cost is high	Computational cost is relatively low

Table 3.1. *Comparison between the implicit and explicit methods*

Figure 3.1 shows the simulation flowchart of the dynamic response and failure of polycarbonate products based on the explicit method.

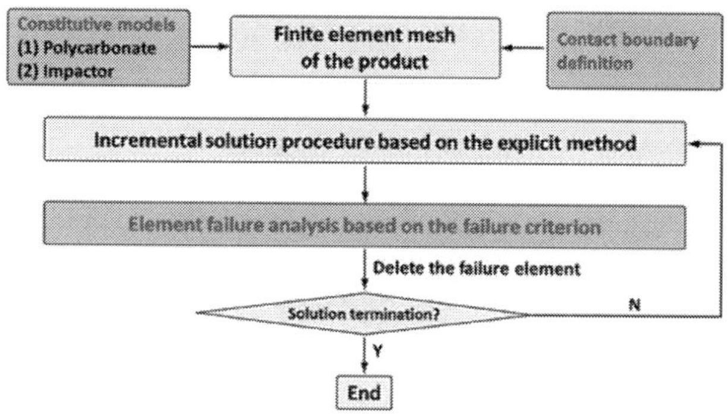

Figure 3.1. *Simulation flowchart for polycarbonate products under impact loading*

3.1.1.1. *Constitutive models of polycarbonate and impactor*

The material properties of polycarbonate can be characterized by the damage-based elastic–viscoplastic constitutive model presented in Chapter 2.

An impactor can generally be studied in two different ways: soft impactor and hard impactor. According to Yankelevsky (1997), a hard impactor is often considered as a projectile that remains undamaged and undergoes small deformation during the impact. In the FE simulation, the rigid model is often adopted to represent the hard impactor (Chen and Li 2004). A soft impactor is commonly defined in such a way that the generated stress far exceeds the projectile strength but is far below the target strength (Wilbeck 1978; Boehman and Challita 1982). For example, a bird is considered as the most representative soft impactor in aeronautic applications. In fact, it is difficult to precisely define the constitutive model of the bird because of its mixed composition of skeleton, blood as well as muscle. It holds an intervening status between liquid and solid during the impact. Two kinds of models, i.e. the elastic–plastic model (Wang *et al.* 2007; Zhu *et al.* 2009; Wang and Yue 2010) and equation of state (EOS) combined with the smooth particle hydrodynamics (SPH) method (Johnson and Holzapfel 2003; Mccarthy *et al.* 2004; Grimaldi *et al.* 2013), have been used to model the bird.

Various elastic–plastic models exist in the material model library of LS-DYNA. For example, the model **MAT_PLASTIC_KINEMATIC* (Hallquist 2007) is generally used to simulate the bird. Table 3.2 lists the parameters of an elastic–plastic model of the bird. The model parameters and the value of failure strain were identified in Wang *et al.* (2007) by the optimization method based on the experimental results of bird strikes on an aluminum panel.

Density (kg/m^3)	Young's modulus (GPa)	Poisson's ratio	Yield stress (MPa)	Tangent modulus (MPa)	Failure strain
950	10	0.3	0.5	3.5	1.22

Table 3.2. *Elastic–plastic model of the bird*

However, the elastic–plastic model cannot accurately capture the bird behavior during the high-speed impact. Experimental studies have shown that the bird behaves like a fluid during the high-speed impact (Abrate 2016). Recently, the equation of state (EOS) has been applied in bird modeling.

The EOS is often used to describe the pressure–volume behavior of fluid. Various versions of the EOS such as polynomial, Murnaghan and Gruneisen forms of equations have been developed in bird modeling. Let us take the most commonly used polynomial EOS as an example. The hydrostatic pressure P of the fluid is formulated as

$$P = C_o + C_1\mu + C_2\mu^2 + C_3\mu^3 \quad [3.3]$$

where

$$\mu = \frac{\rho}{\rho_o} - 1, \quad [3.4]$$

C_o, C_1, C_2 and C_3 are material constants and μ is the ratio of the current density ρ to the initial density ρ_0. The properties and polynomial EOS parameters of the bird (Grimaldi *et al.* 2013) are provided in Table 3.3.

Density (kg/m^3)	**Shear modulus (GPa)**	**Yield stress (MPa)**	**EOS parameters**			
			C_o	C_1	C_2	C_3
950	2	0.02	0	2.1×10^9	6.2×10^9	10.1×10^9

Table 3.3. *Properties and polynomial EOS parameters of the bird*

EOS is often combined with the smooth particle hydrodynamics (SPH) method to model the highly deformable bird. The Lagrangian or Eulerian descriptions in the mesh-based methods have difficulties in dealing with large deformations due to severe mesh distortion. The meshless SPH method is devised to consider the material as a series of particles. It is considered as a suitable approach for modeling a highly deformable bird. In the SPH method, the bird is considered as a set of discrete interacting particles. The field variable of any discrete particle can be computed through the interpolation of neighboring particles. At present, the SPH method is implemented in various commercial software such as LS-DYNA and ABAQUS (Mccarthy *et al.* 2004). The basic principles of the SPH method are presented in (Belytschko *et al.* 2004; Abrate 2016).

3.1.1.2. *Contact boundary definition*

Contact impact is one of the most difficult nonlinear problems in engineering because mechanical responses are discontinuous. LS-DYNA provides several contact models that can simulate the majority of contact boundaries. For the impact of polycarbonate products, the following contact models are recommended:

(a) **CONTACT_NODES_TO_SURFACE*;

(b) **CONTACT_AUTOMATIC_NODES_TO_SURFACE*;

(c) **CONTACT_ERODING_NODES_TO_SURFACE*;

(d) **CONTACT_SURFACE_TO_SURFACE*;

(e) **CONTACT_AUTOMATIC_SURFACE_TO_SURFACE*;

(f) **CONTACT_ERODING_SURFACE_TO_SURFACE.*

In the present study, **CONTACT_ERODING_NODES_TO_SURFACE* is adopted in the simulation. The keywords and the corresponding parameters are defined as follows:

**CONTACT_ERODING_NODES_TO_SURFACE*

SSID	MSID	SSTYP	MSTYP	SBOXID	MBOXID	SPR	MPR
1	2	3	3	Optional	Optional	0	0
FS	FD	DC	VC	VDC	PENCHK	BT	DT
0.15	0.55	0.0	0.0	0.0	0.0	0.0	1.0E20
SFS	SFM	SST	MST	SFST	SFMT	FSF	VSF
1.0	1.0	Element thickness	Element thickness	1.0	1.0	1.0	1.0
ISYM	EROSOP	IADJ					
1	1	0					

**CONTROL_CONTACT*

SLSFAC	RWPNAL	ISLCHK	SHLTHK	PENOPT	THKCHG	ORIEN
1.0	0.0	2	0	1	1	1
USRSTR	USRFAC	NSBCS	INTERN	XPENEN		
0	0	10	0	4.0		

Among these parameters, SSID and MSID are slave and master segments respectively that determine the contact searching direction. FS and FD are static and dynamic coefficients of friction respectively. SFS and SFM are scale factors on slave and master penalty stiffness respectively. EROSOP verifies whether interior node information is saved when eroding contact occurs. ISLCHK checks the initial penetration. The details of the contact model and the corresponding parameters can be found in the LS-DYNA Keyword User's Manual (Hallquist 2007).

3.1.1.3. *Failure criterion*

According to the plastic deforming ability before the fracture, two failure modes are defined: ductile and brittle failure modes. In ductile failure, the crack grows at a slow pace and is accompanied by extensive plastic deformation and energy absorption. In contrast, brittle failure is characterized by rapid crack propagation with low energy absorption but without significant plastic deformation. Figure 3.2 shows the schematic representation of different failure modes.

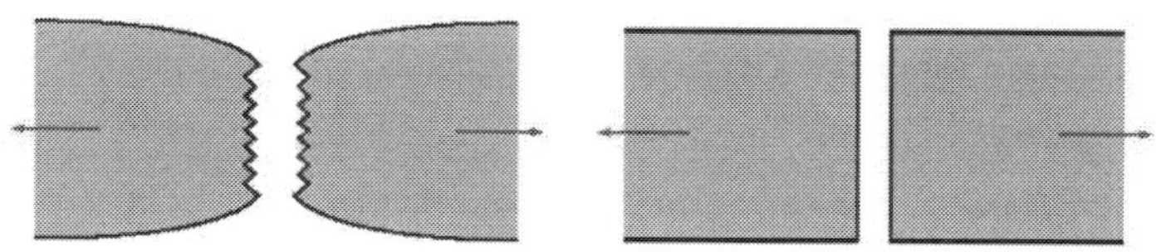

(a) Ductile failure (b) Brittle failure

Figure 3.2. *Schematic representation of different failure modes*

Most metals and polymers undergo ductile failure at room or higher temperatures. Only ceramics or metals exhibit typical brittle failure at very low temperatures. Thus, various criteria have been proposed to model the ductile failure, such as the equivalent plastic strain, maximum shear stress and Johnson–Cook. The equivalent plastic strain criterion is the simplest and therefore one of the most widely used to model the failure of polycarbonate. It is formulated as

$$\mathcal{G} = \sqrt{\frac{2}{3}\boldsymbol{\varepsilon}_p : \boldsymbol{\varepsilon}_p} - \varepsilon_f \qquad [3.5]$$

where $\boldsymbol{\varepsilon}_p$ is the plastic strain tensor and ε_f denotes the ductile failure strain. The material failure occurs if the fracture function $\mathcal{G} = 0$. Otherwise, the material is within the safe domain for $\mathcal{G} < 0$. The failure strain of polycarbonate at room temperature is estimated to be about 1.35 based on the experimental observations (Duan *et al.* 2003).

In LS-DYNA, the *ADD_EROSION* option provides a convenient way to include the failure criterion in the model. Once the failure criterion defined in the keyword **MAT_ ADD_EROSION* is satisfied, the element is deleted from the calculation. The keywords and the corresponding parameters are defined as follows:

MID	EXCL						
1	123.0						
PFAIL	SIGP1	SIGVM	EPSP1	EPSSH	SIGTH	IMPILSE	FAILTM
123.0	123.0	123.0	1.35	123.0	123.0	123.0	123.0

where MID is the material number to which the failure criterion is applied. PFAIL to FAILTM are different failure models (for details, see the LS-DYNA Keyword User's Manual (Hallquist 2007)). In the failure criterion in equation [3.5], the equivalent plastic strain failure model EPSP1 is used. When the maximum equivalent plastic strain reaches the failure strain, the element deletion will be activated.

3.1.2. *Gas gun impact tests*

The proposed impact simulation methodology is verified by comparing with the experimental results. High-velocity impacts on polycarbonate plates are carried out by using a gas gun impact test. During testing, the steel bullet accelerated by the gas gun system strikes against the center of the square polycarbonate plate. The material of the bullet is steel C45E4 (ISO) and the mass is 38.7 g. Geometries and dimensions of the bullet and the polycarbonate plate are shown in Figure 3.3.

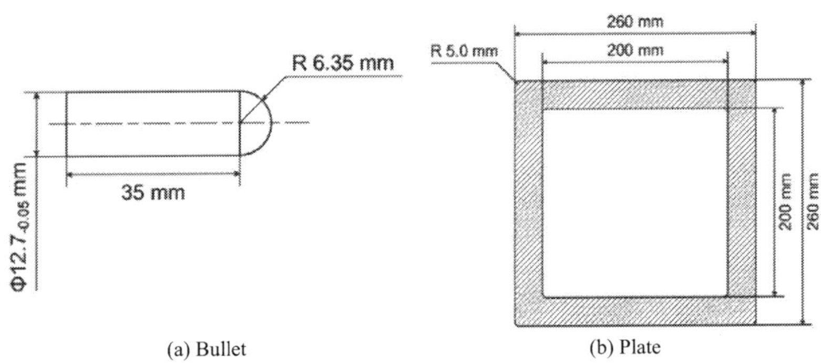

Figure 3.3. *Geometries and dimensions of the bullet and the polycarbonate plate*

(a) Layout

(b) High-speed camera

(c) Laser speedometer

(d) Fixture and platform

(e) Data acquisition system

(f) Schematic representation

Figure 3.4. *Photographs of the gas gun impact test setup*

The experimental setup consists of an air gun system, a fixture and platform, a high-speed camera, a laser speedometer and a data acquisition system. Photographs and their schematic representation are shown in Figure 3.4. The shaded area on the plate in Figure 3.3(b) is clamped by the fixture, the velocity of the bullet is measured by the laser speedometer and the entire impact process is recorded by the high-speed camera.

In the tests, different thicknesses of the plate and velocities of the bullet are considered. For the plate with a thickness of 1.0 mm, the impact tests are performed with bullet velocities of 46.16, 49.75 and 53.69 m/s. For a thickness of 1.5 mm, the bullet velocities are 62.58, 64.24 and 65.33 m/s. For a thickness of 2.0 mm, the bullet velocities are 68.26, 68.78 and 69.55 m/s. For a thicknesses of 3.0 and 5.0 mm, the bullet velocities are 79.30 and 82.08 m/s respectively.

Experimental results are shown in Figure 3.5. Two phenomena are observed on the polycarbonate plates in the impact tests: crater and perforation. For the three thin plates with thicknesses of 1.0, 1.5 and 2.0 mm, the critical perforation velocities are about 49.75, 64.24 and 68.78 m/s respectively. For the two thick plates with thicknesses of 3.0 and 5.0 mm, the maximum velocity generated by the gas gun system is not high enough to perforate the plates so that only hemispherical craters are observed.

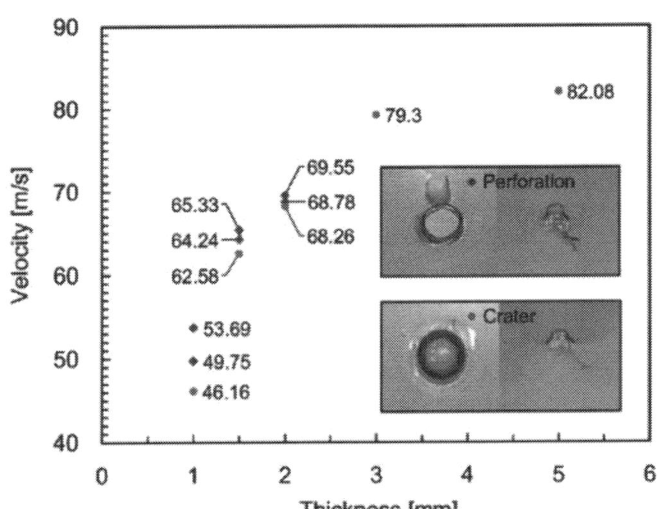

Figure 3.5. *Experimental results of the gas gun impact tests. For a color version of this figure, see www.iste.co.uk/zhang/polycarbonate.zip*

3.1.3. *Gas gun impact simulations*

In this section, gas gun impact tests on polycarbonate plates are simulated. FE models of the bullet and plate are shown in Figure 3.6. Both the bullet and plate are meshed using eight-node explicit 3D elements SOLID164 in LS-DYNA. To precisely capture the dynamic response around the impact points, a finer mesh is used in the region enclosed by a dashed red line, as shown in Figure 3.6(b). In the simulations, the bullet strikes against the center of the plate with an initial velocity while the nodes on the four sides of the plate are clamped.

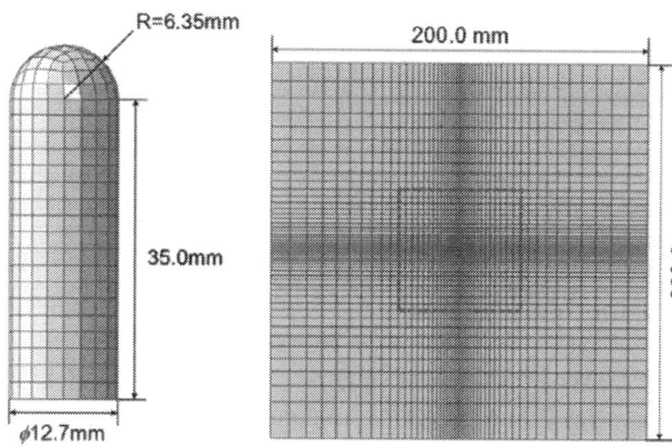

(a) Bullet (b) Plate

Figure 3.6. *FE models in the gas gun impact simulation. For a color version of this figure, see www.iste.co.uk/zhang/polycarbonate.zip*

The bullet is assumed to be rigid with a Young's modulus of 200 GPa and a Poisson's ratio of 0.3. The polycarbonate plate is represented by the elastic–viscoplastic constitutive model developed in Chapter 2. The contact model provided in LS-DYNA,**CONTACT_ERODING_NODES_TO_SURFACE* is used to simulate contact between the bullet and the plate during the impact. The ductile failure criterion given in equation [3.5] is defined through the keyword **MAT_ADD_EROSION* in order to capture the crater and perforation on the polycarbonate plate.

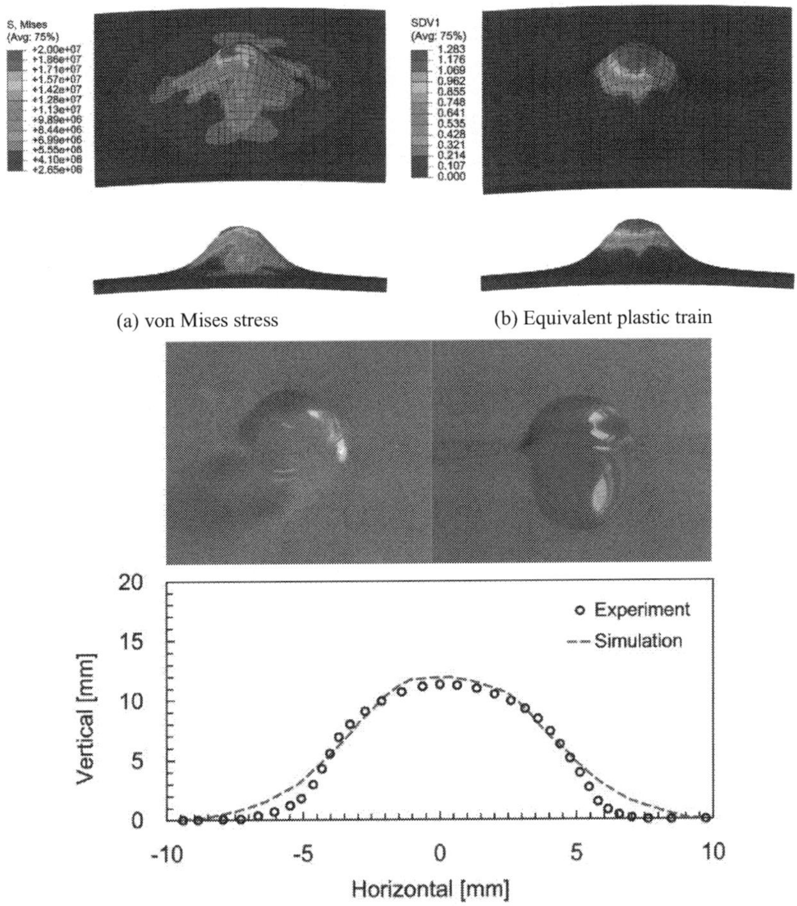

Figure 3.7. *Comparison of the simulation result and experiment. For a color version of this figure, see www.iste.co.uk/zhang/polycarbonate.zip*

Figure 3.7 shows the simulation result and the experimental photograph of the impact test with a plate thickness of 2.0 mm and a bullet velocity of 68.26 m/s. In order to highlight the response around the impact points, numerical results within the enclosed region are compared. The contours of von Mises stress and the equivalent plastic strain on the plate are shown in Figure 3.7(a) and (b) respectively. The maximum values are located on the impact points. Figure 3.7(c) shows the experimental photographs of the crater on the plate, the simulated profile of the impacted plate and the experimentally measured profile with good agreement.

t = 1.0 mm v = 53.69 m/s	t = 5.0 mm v = 82.08 m/s
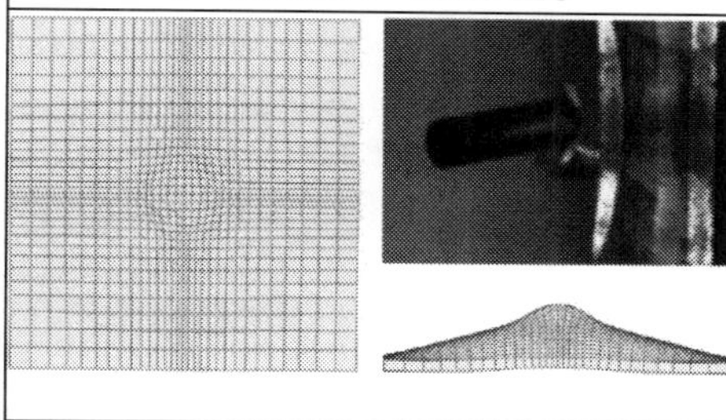	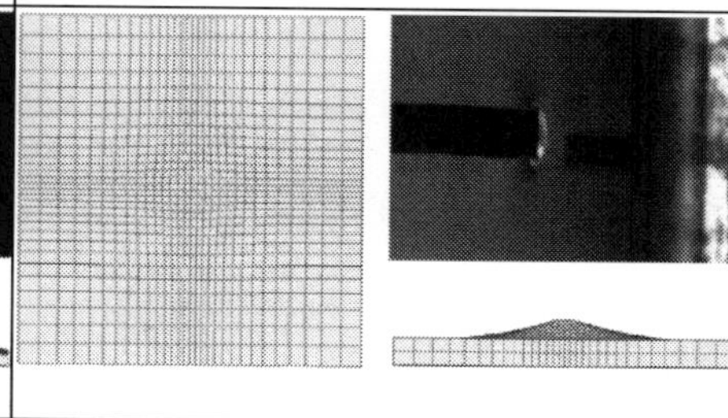
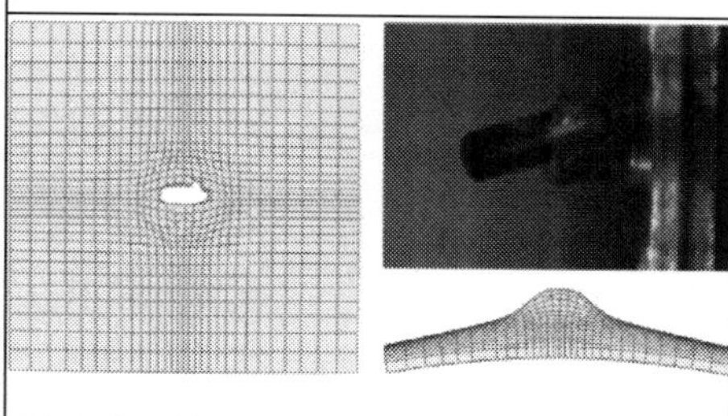	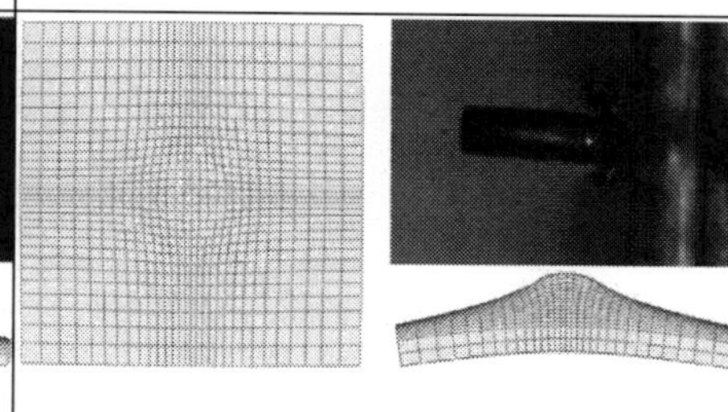
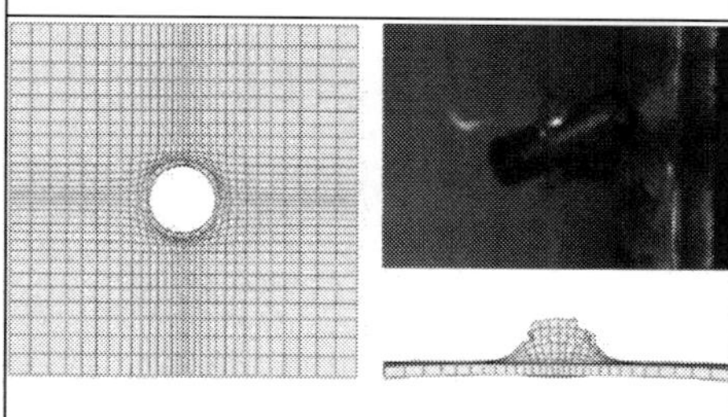	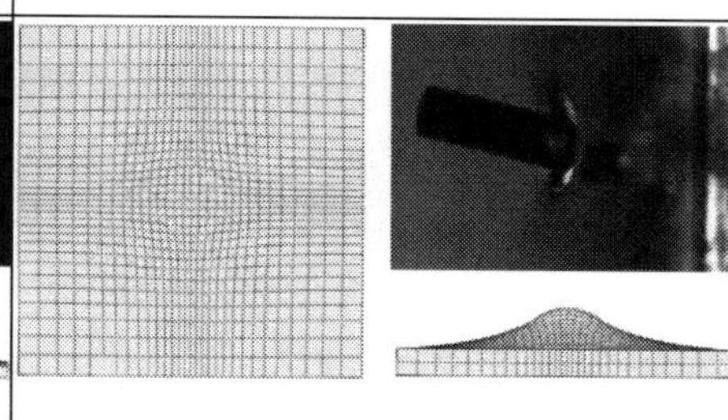
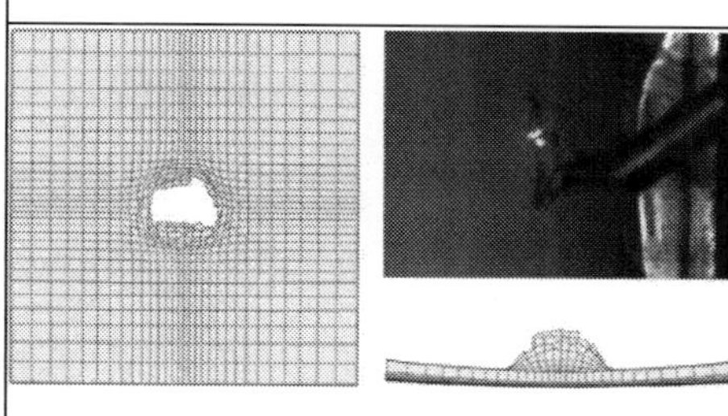	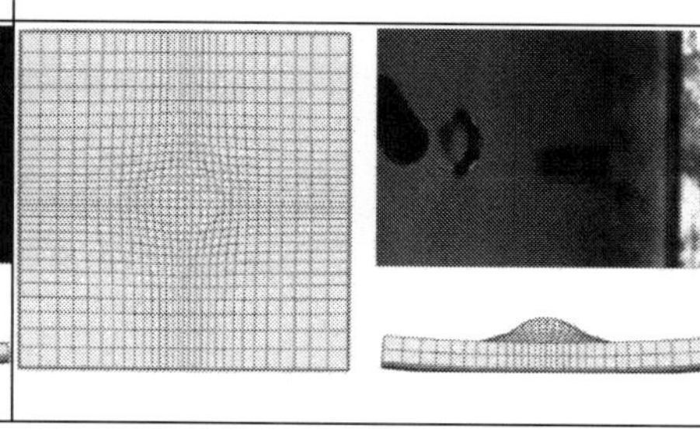

Table 3.4. *Comparisons of the simulation results and the experimental photographs for the gas gun impact tests*

To further investigate the fracture of the polycarbonate plate under high-velocity impacts, two other impact simulations with different plate thicknesses and bullet velocities are performed. Comparisons of the simulation results and the experimental photographs are given in Table 3.4. In the first test, the plate thickness and the bullet velocity are 1.0 mm and

53.69 m/s respectively. In the second test, they are 5.0 mm and 82.08 m/s respectively. The plate in the first test undertakes roughly four stages: protrusion, crack, penetration and perforation. Finally, an irregular opening is left on the plate. In the second test, the bullet does not perforate the plate and only leaves a crater on the plate. A good correlation between the simulation results and the experimental photographs validates the proposed impact simulation methodology for the deformation and fracture of the polycarbonate plate under high-velocity impacts.

3.2. Impact simulation in aeronautical and aerospace applications

3.2.1. *Polycarbonate windshield under bird impact*

Bird impact is regarded as a serious threat to both military and civilian aircrafts (Thorpe 1997; Cleary 2006; Doubrava and Strnad 2010). Exposed aircraft components such as windshields, leading edges, engine structures and blades are vulnerable to bird strikes. When fighter aircrafts are operated at low altitude and high speed, the probability of bird impact increases greatly and proves to be lethal to the safety of the pilot and aircraft components. The average impact force imparted by a 1.8 kg bird to an aircraft traveling at 600 km/h is approximately 165 kN (18 tons).

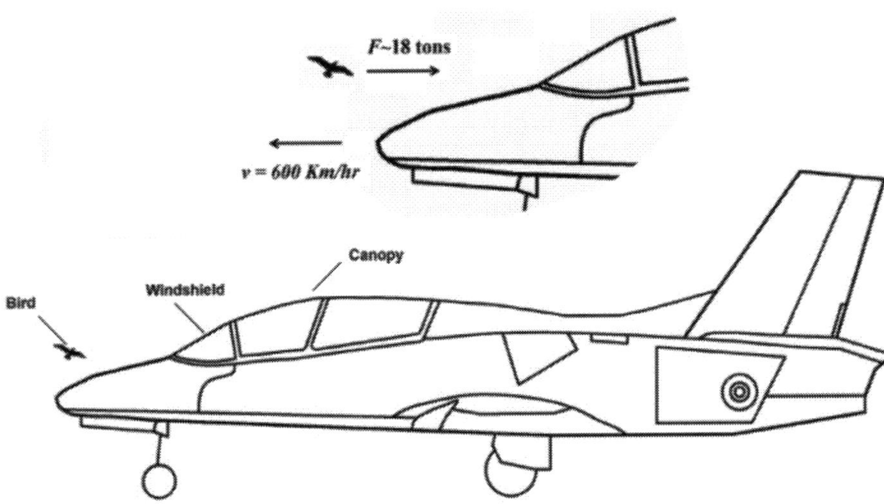

Figure 3.8. *Windshield and canopy of military aircraft*

Figure 3.8 presents a lateral view of a military aircraft showing possible bird impacts at the windshield or canopy. The windshield is the frontal part of the aircraft which is susceptible to bird impact. In order to ensure the safety of aircraft, the windshield must be capable of withstanding high-velocity impact threats. Design of an impact-resistant windshield is challenging and requires extensive experimental testing conducted in the past. The FE method is adopted as an effective tool to simulate the bird strike. It allows us to analyze the most stringent impact conditions that cannot be considered in experiments, and provides detailed insights into the impact process difficult to observe during experimental testing.

3.2.1.1. *FE modeling of windshield structure*

The whole windshield structure consists of a polycarbonate windshield, a rubber gasket and an aluminum alloy frame. The corresponding FE model consists of 48,600 SOLID 164 elements of LS-DYNA, as shown in Figure 3.9.

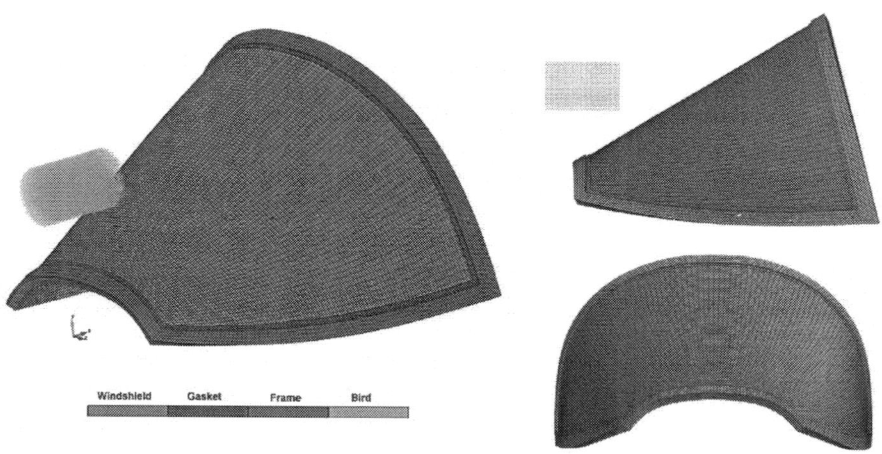

Figure 3.9. *FE model for bird impact on the windshield. For a color version of this figure, see www.iste.co.uk/zhang/polycarbonate.zip*

Polycarbonate is represented by the elastic–viscoplastic constitutive model. The aluminum frame is modeled as a plastic material using the material model **MAT_PLASTIC_KINEMATIC* in LS-DYNA. The rubber gasket between the frame and the windshield is modeled with the

Mooney–Rivlin hyperplastic material model **MAT_MOONEY-RIVLIN _RUBBER* in LS-DYNA. Material parameters for the gasket and the frame are listed in Table 3.5.

Aluminum alloy frame					
Density *($kg.m^{-3}$)*	Elastic modulus *(GPa)*	Tangent modulus *(MPa)*	Poisson's ratio	Yield strength *(MPa)*	Failure strain
2,780	72	690	0.3	313	0.28

Mooney–Rivlin parameter for rubber				
Density *($kg.m^{-3}$)*	Elastic modulus *(MPa)*	Poisson's ratio	A *(MPa)*	B *(MPa)*
1,100	310	0.49	0.378	0.977

Table 3.5. *Material parameters of the surrounding frame and gasket*

3.2.1.2. *SPH method and polynomial EOS for bird modeling*

The SPH method and polynomial EOS are combined to model the bird in the present case. The properties and polynomial EOS parameters of the bird are listed in Table 3.3. Straight cylinders have been proved to be the best shape to simulate the impacts of real birds (Barber *et al.* 1978), and a length-to-diameter ratio of 1.5–2 is recommended in (Challita 1980). Dimensions of the bird can be determined from its mass and density. Commercial aircraft components should withstand the strike of a bird weighing up to 1.8 kg (Liu *et al.* 2014). In this study, a 1.8 kg bird is modeled as a cylinder of 120 mm in diameter and 180 mm in length, as shown in Figure 3.10. The SPH model of the bird consists of 32,220 nodes. Each node has a 0.0557 g lumped mass, and the average distance between two neighboring nodes is 4 mm.

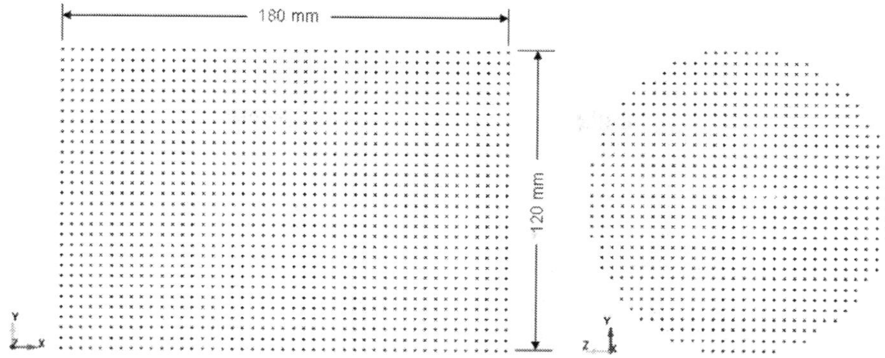

Figure 3.10. *SPH model of the bird*

3.2.1.3. *Experimental verification*

In the experimental studies (Liu *et al.* 2014), the polycarbonate windshield impacted by a 1.8 kg bird with a velocity of 64.4 m/s at four different locations (1, 2, 3 and 4) is shown in Figure 3.11. The displacement profiles at four locations were recorded during the experiment.

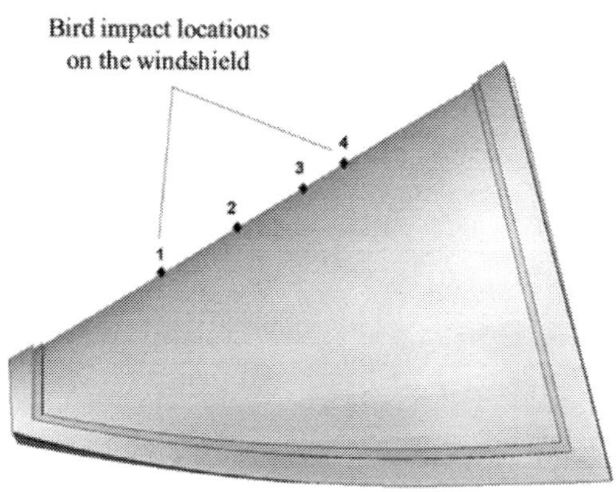

Figure 3.11. *Bird impact at four different positions on the windshield*

Let us assume that the edge of the windshield is fully fixed in the FE simulation. The ductile failure criterion is defined by **MAT_ADD_EROSION* to capture the failure of polycarbonate windshield.

The contact model **CONTACT_ERODING_NODES_TO_SURFACE* is used to model the interaction between the bird and the windshield.

Simulation results and experimental data are compared in Figure 3.12. The simulated displacement curves at the four locations are found to be in good agreement with the experiments.

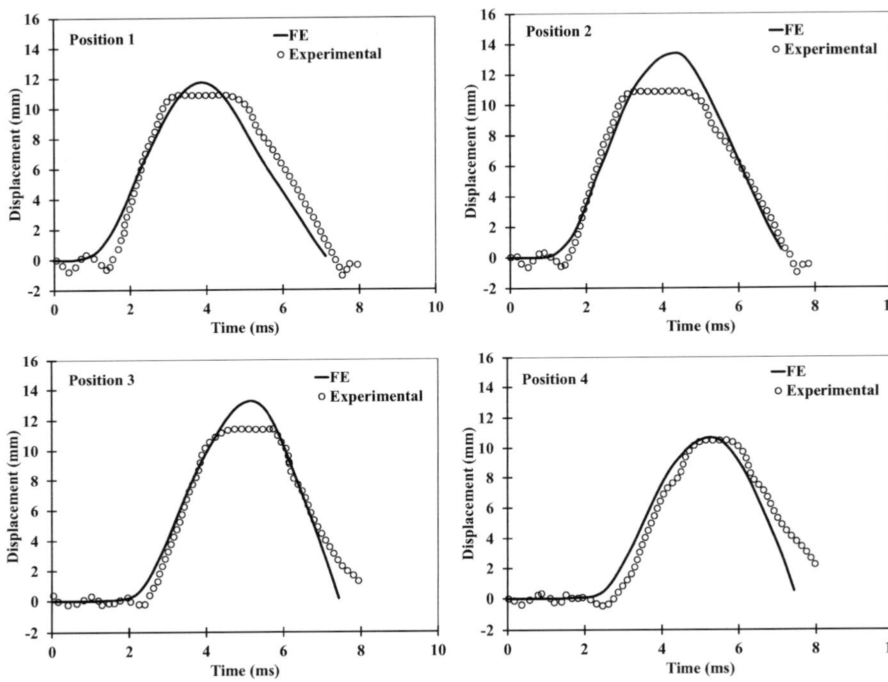

Figure 3.12. *Comparisons of simulation results with experimental data*

3.2.1.4. *Simulations at various impact velocities, impact angles and temperatures*

FE simulations are performed over a range of impact velocities when the bird impacts at the central point of the windshield. With the increase in bird impact velocity, the deformations of the windshield and related components increase due to the increasing kinetic energy of the bird and its high impact force. Figure 3.13 shows the sequence of impacts at various time intervals with an initial bird velocity of 65 m/s. The whole impact event lasts 14 ms.

The equivalent stress fields in different windshield components are shown correspondingly. It can be seen that the surrounding frame of the windshield remains in the state of high stress during the impact. The bird crumbles and slides along the surface of the windshield.

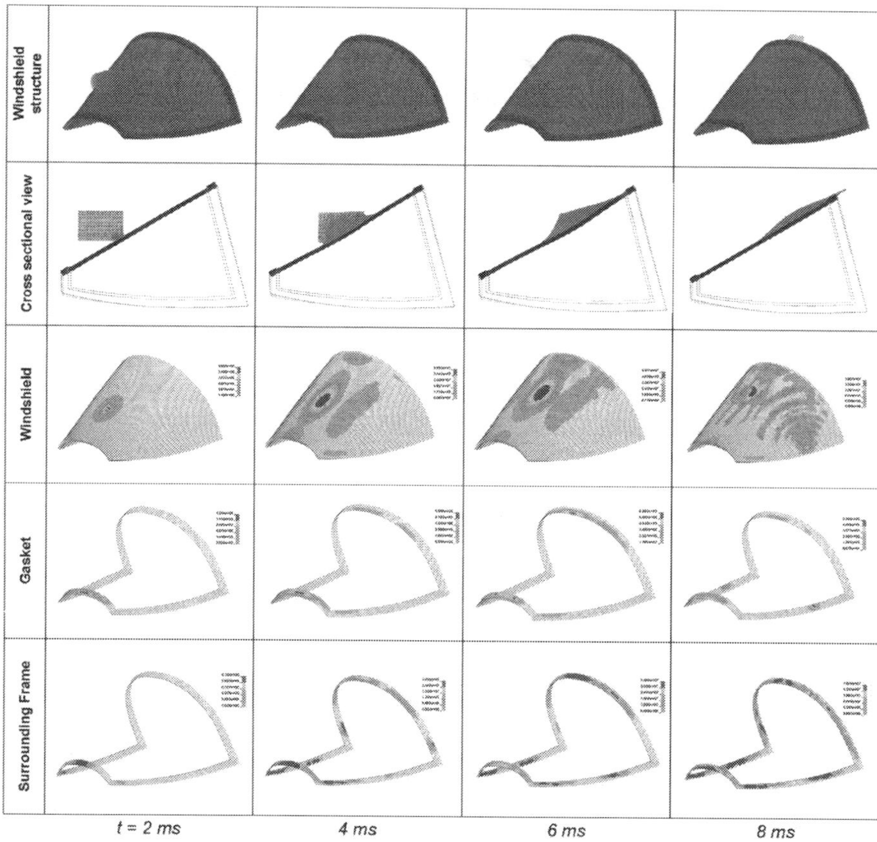

Figure 3.13. *Bird impact on the windshield structure at different time intervals. For a color version of this figure, see www.iste.co.uk/zhang/polycarbonate.zip*

Figure 3.14 indicates that the resultant normal displacement at the impact point increases along with the impact velocity, while the time to reach the maximum displacement decreases with the increase in impact velocity.

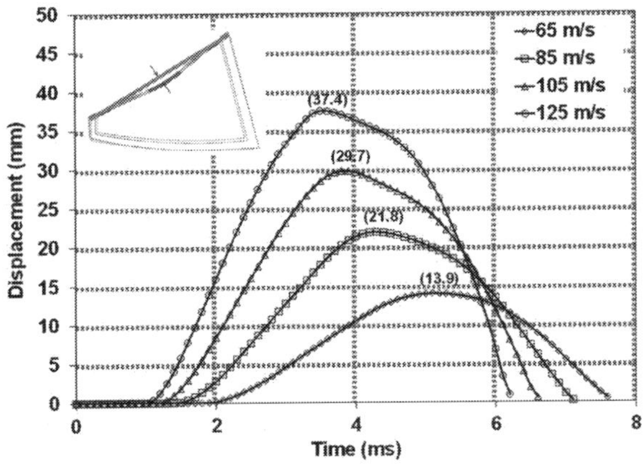

Figure 3.14. *Normal displacement at the central impact point of the windshield. For a color version of this figure, see www.iste.co.uk/zhang/polycarbonate.zip*

Figure 3.15 shows that the magnitude of equivalent stress also increases with the increase in impact velocity. At the impact velocity of 65 m/s, the value of maximum stress is 38.7 MPa, which increases to 69.5 MPa when the impact velocity attains 125 m/s.

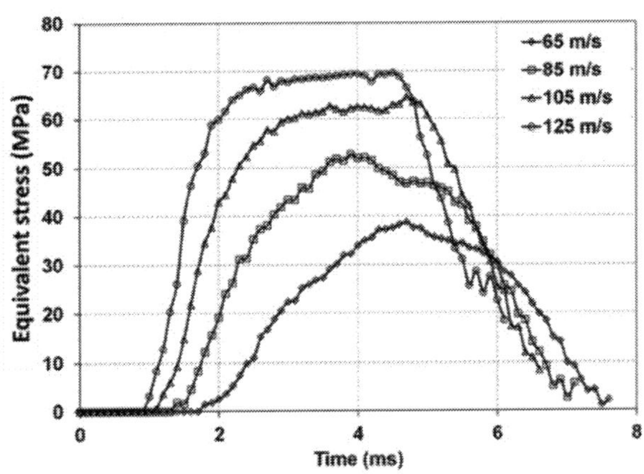

Figure 3.15. *Equivalent stress history at the central impact point of the windshield*

In summary, the maximum values of the normal displacement and equivalent stress at different locations along the central camber line of the windshield are plotted in Figure 3.16. It can be seen that the increasing velocity enlarges the deformation and equivalent stress at the upper half of the windshield.

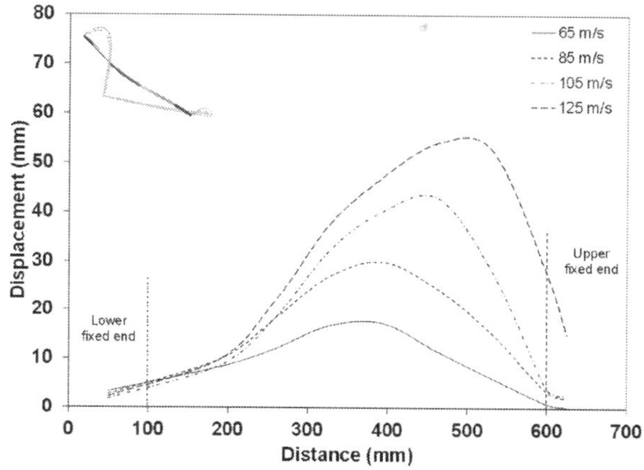

(a) Normal displacement

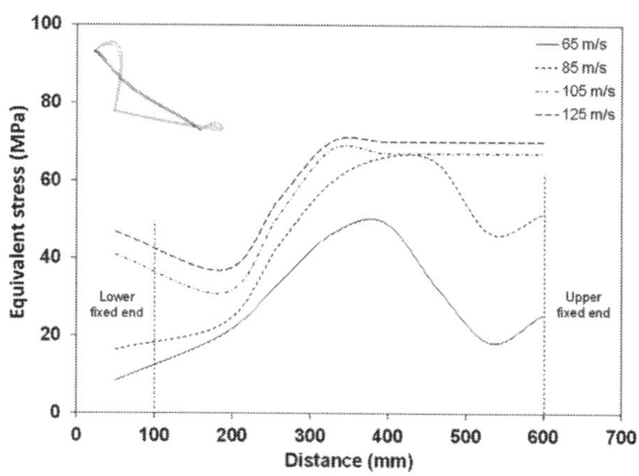

(b) Equivalent stress

Figure 3.16. *Normal displacement and equivalent stress along the central line. For a color version of this figure, see www.iste.co.uk/zhang/polycarbonate.zip*

This implies that the upper fixed end of the windshield is susceptible to failure. Figure 3.17 further shows the normal displacement and equivalent stress at the upper fixed end of the windshield when the bird impacts at the central point with a velocity up to 175 m/s. In this case, the upper fixed end of the windshield fails. The critical impact velocity is 133.4 m/s, and the critical impact energy is 16 kJ.

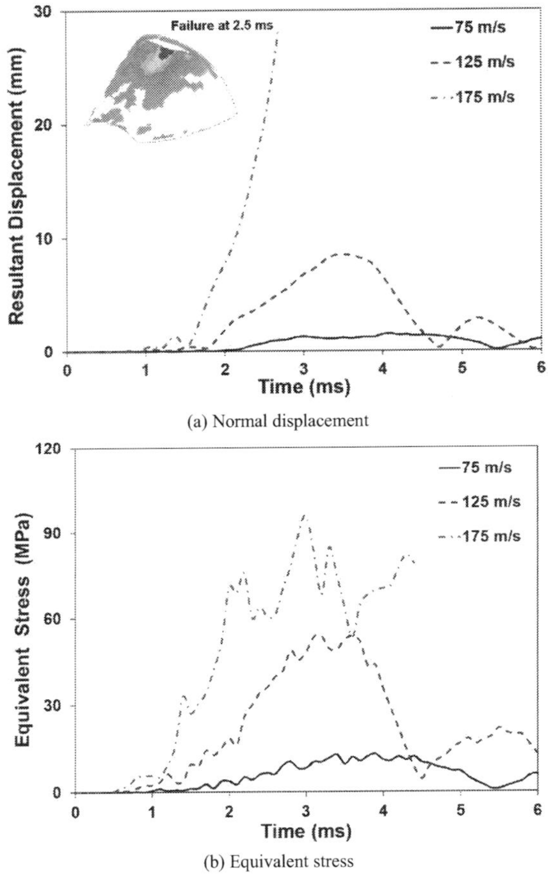

Figure 3.17. *Normal displacement and equivalent stress at the upper fixed end of the windshield. For a color version of this figure, see www.iste.co.uk/zhang/polycarbonate.zip*

The influence of the bird impact angle is also studied. Three different angles (i.e. 15°, −15° and 0°) with respect to the horizontal directions are considered, as shown in Figure 3.18. Note that the 0° impact is not shown in the figure.

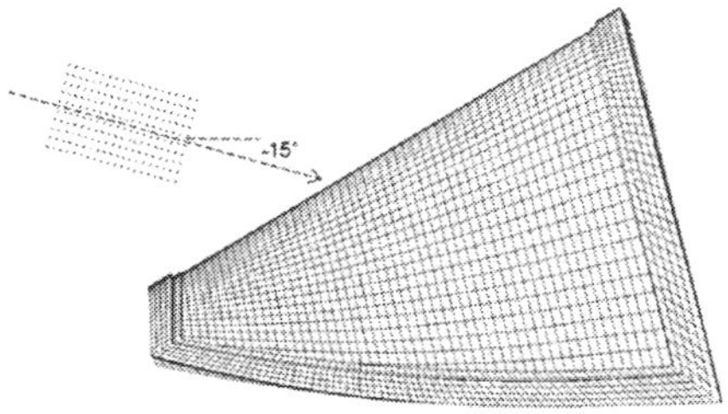

(a) -15°

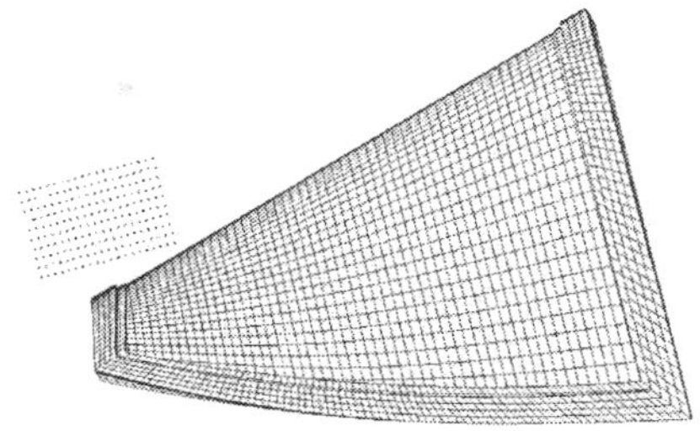

(b) 15°

Figure 3.18. *Bird impact on the windshield at different angles*

Let us suppose that the impact velocity is 65 m/s. The corresponding normal displacement, stress and impact force at the impact point of the windshield are shown in Figure 3.19. At the impact angle of 15°, the bird slightly fails and slides along the windshield surface. At the impact angle of −15°, the bird transfers most of its energy to the windshield so that high displacement and stress occurs at the impact point. Most of the bird body fails during the impact. In Figure 3.19, it can be observed that at the impact angle of 15°, the maximum normal displacement is 3.85 mm, which increases to 13.95 and 22.89 mm when the impact angle is changed to 0° and −15° respectively. The impact force also sharply increases with the change in the impact angle. At 15°, the maximum impact force is 10.2 kN, which rises to 44.1 kN for the impact angle of −15°. The peak impact force increases almost 4.3 times when the impact angle is changed from 15° to −15°.

Moreover, it can be seen that at the impact angle of 15°, the maximum stress level is 11 MPa. When the impact angle changes to −15°, the peak stress value dramatically rises to 62 MPa, which approaches the yield stress of polycarbonate. Therefore, the impact angle is a critical parameter in determining the critical impact velocity of the windshield.

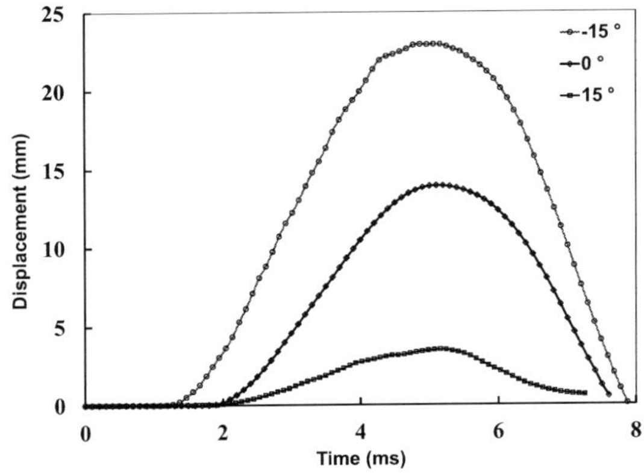

(a) Normal displacement

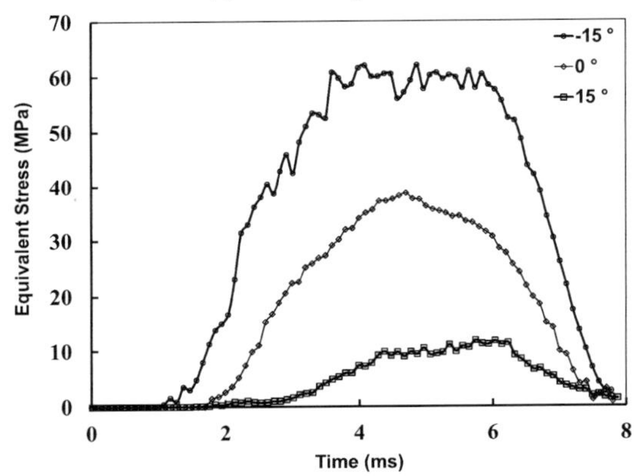

(b) Equivalent stress

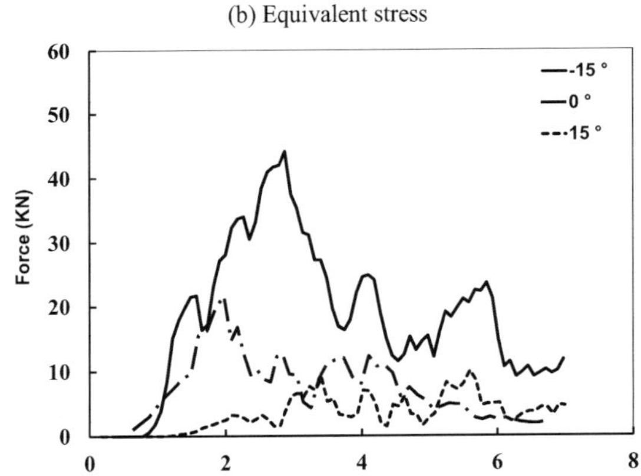

(c) Impact force

Figure 3.19. *Displacement, stress and impact force histories at the impact point of the windshield at various impact angles*

The temperature effect is also investigated. Figure 3.20 shows the displacement–time plots at the central impact point of the windshield at the bird impact velocity of 65 m/s. It concludes that the maximum normal displacement increases with the increase in temperature. When the temperature increases from 313 to 373 K, the maximum displacement increases up to two times, but the equivalent stress decreases. That is, the maximum stress decreases to about 32%.

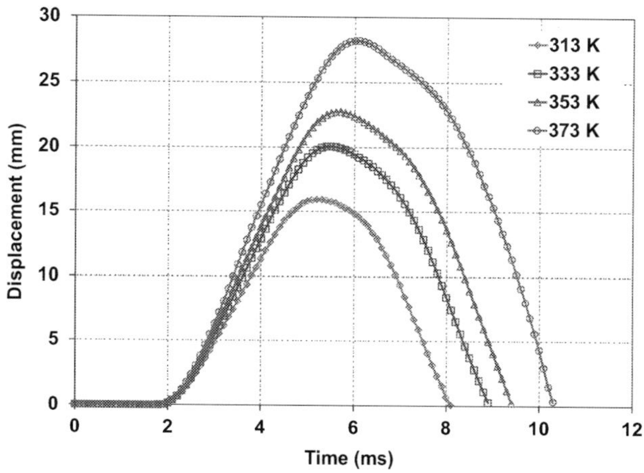

(a) Normal displacement

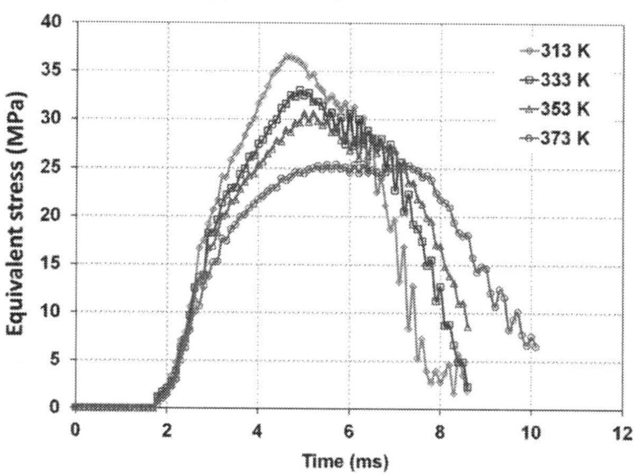

(b) Equivalent stress

Figure 3.20. *Temperature effect on normal displacement and equivalent stress at the central impact point of the windshield. For a color version of this figure, see www.iste.co.uk/zhang/polycarbonate.zip*

3.2.2. *Polycarbonate visor under debris impact*

The optical visor is an important part of the astronaut's helmet. As shown in Figure 3.21, the polycarbonate visor provides the astronaut with not only a clear view, but also with protection against harmful radiations and debris impact. The impact threat to the astronaut is mainly caused by the small-size debris that cannot be eye-tracked. In this section, FE simulation is used to investigate the impact behavior and the critical impact energy of a polycarbonate visor.

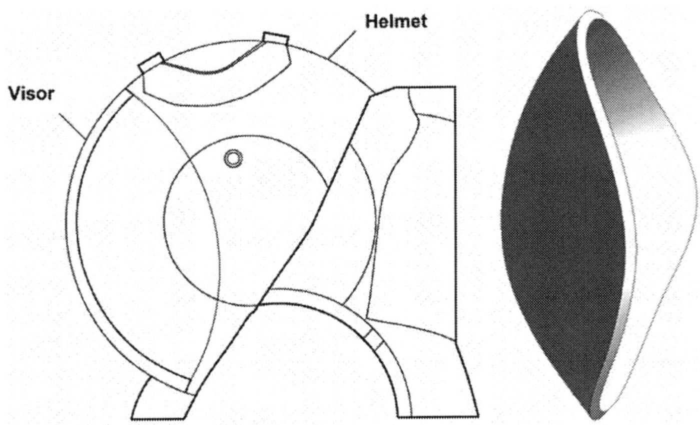

Figure 3.21. *Astronaut's helmet and visor*

Let us suppose that a spherical debris with a diameter of 10 mm impacts at the central point of the visor. The edge of the visor is fully constrained. The FE models of both the spherical projectile and the visor are presented in Figure 3.22. The visor is meshed with 43,200 SOLID 164 elements in LS-DYNA, while the spherical projectile is meshed with 28,000 SOLID 64 elements.

The projectile is assumed to be rigid with Young's modulus of 200 GPa and Poisson's ratio of 0.3. Likewise, the elastic–viscoplastic constitutive model developed in the previous chapter is used to model the polycarbonate visor. The contact between the projectile and the visor is simulated by the contact model **CONTACT_ERODING_NODES_TO_SURFACE*. The ductile failure criterion is defined by **MAT_ ADD_EROSION* to capture the crater and perforation of the polycarbonate visor.

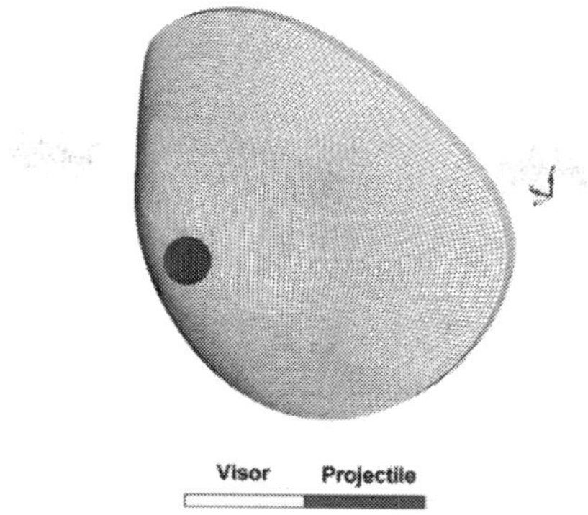

Figure 3.22. *FE models of the projectile and the visor. For a color version of this figure, see www.iste.co.uk/zhang/polycarbonate.zip*

Simulation results of deformation and equivalent stresses of the visor with an impact velocity of 60 m/s at different time intervals are shown in Figure 3.23. The cross-sectional view of the visor is shown to highlight the deformation.

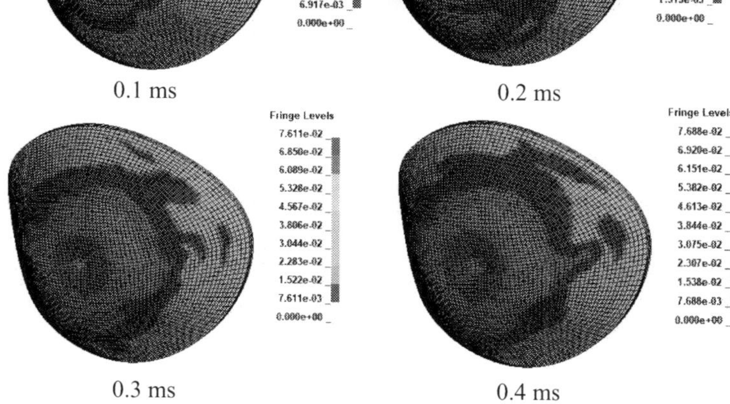

Figure 3.23. *Simulation results for the projectile velocity of 60 m/s at different time intervals. For a color version of this figure, see www.iste.co.uk/zhang/polycarbonate.zip*

With the increase in impact velocity, the normal displacement and stress level in the visor increase dramatically, as shown in Figure 3.24. At the impact velocity of 80 m/s, the visor reaches its yield stress limit and permanent deformation appears at the impact location. At the impact velocity of 120 m/s, the instantaneous failure in the visor is observed at the impact point.

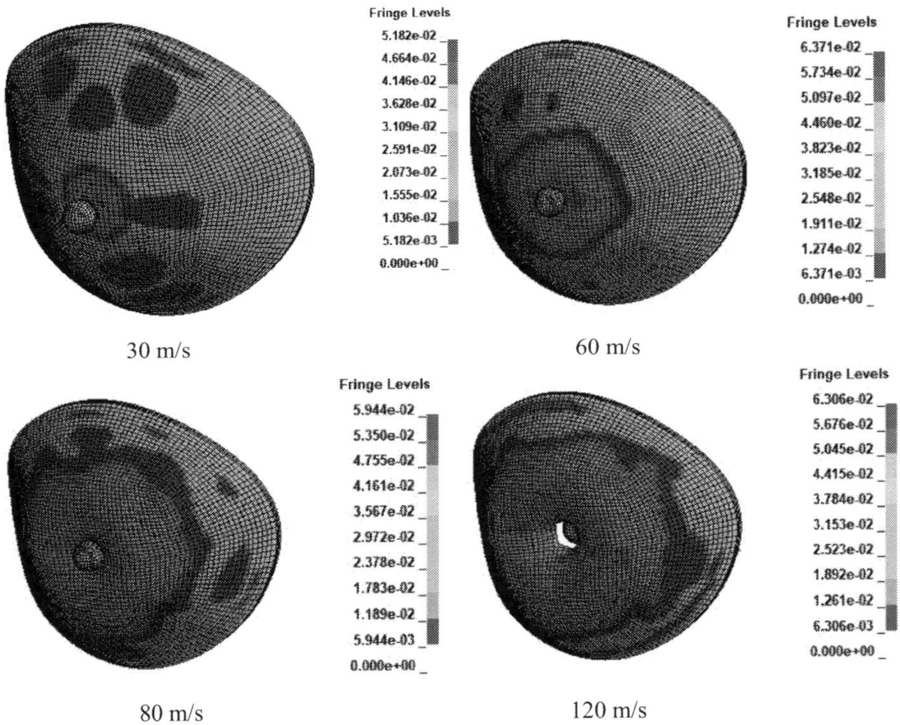

Figure 3.24. *Simulation results at different impact velocities. For a color version of this figure, see www.iste.co.uk/zhang/polycarbonate.zip*

Figures 3.25 and 3.26 plot normal displacement and equivalent stress histories at the impact point at different projectile velocities. With the increase in projectile velocity, the displacement increases at the impact point. The visor fails at the impact velocity of 120 m/s.

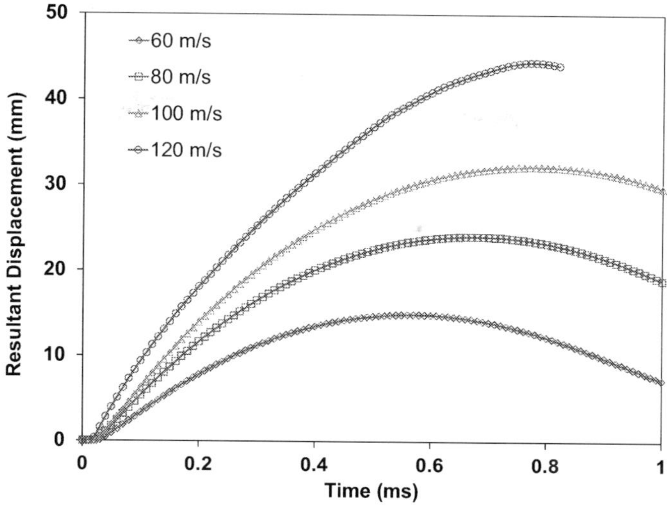

Figure 3.25. *Resultant displacement at the impact point. For a color version of this figure, see www.iste.co.uk/zhang/polycarbonate.zip*

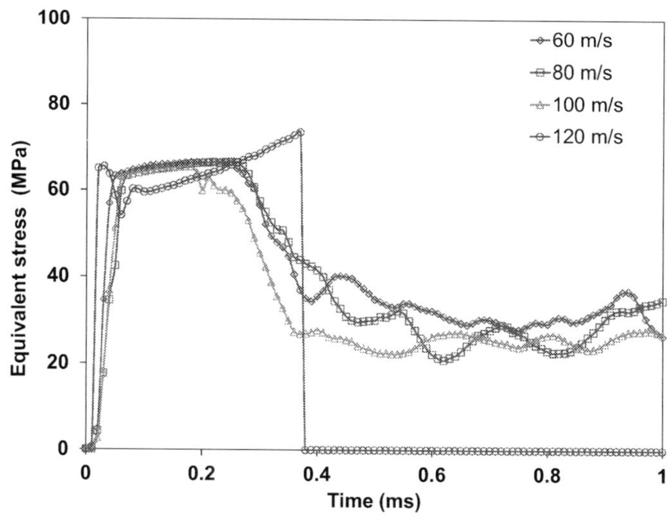

Figure 3.26. *Equivalent stress history at the impact point. For a color version of this figure, see www.iste.co.uk/zhang/polycarbonate.zip*

The variation of residual energy of the projectile at different impact velocities is shown in Figure 3.27. At the impact velocity of 120 m/s, the projectile has enough energy to perforate the visor at the impact point. The minimum energy required for the complete perforation of the visor is 216 J, which is obtained by the subtraction of residual energies of the projectile.

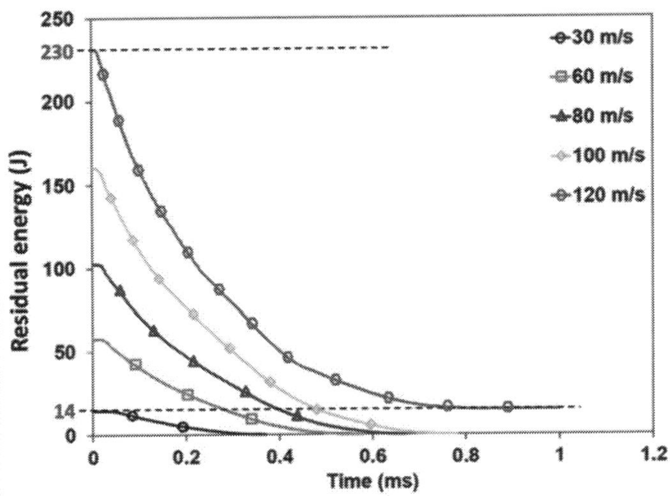

Figure 3.27. *Residual energy of the projectile. For a color version of this figure, see www.iste.co.uk/zhang/polycarbonate.zip*

3.3. Conclusion

In this chapter, a simulation methodology was presented to predict the high-speed impact behavior of polycarbonate products. Basic problems such as the material model, the impact–target contact definition, the failure criteria and the implementation in LS-DYNA were discussed in detail.

First, simulations of the gas gun impact test of polycarbonate plates were carried out and validated against experimental data. The comparison with experimental results showed the validity of the proposed methodology. Then, dynamic responses of the windshield against bird impact and the astronaut's helmet visor against debris impact were simulated, and the effects of various parameters on their dynamic responses were studied. Impact velocity is one of the most critical parameters that strongly influenced the dynamic responses. Numerical models were able to predict the impact responses of both structures for a range of impact velocities. On the basis of the simulation results, the critical velocity and weak locations of the structures were determined. For the range of impact velocities considered in this study, the admissible impact energies for the safe operation of the windshield and visor were also determined. In addition, the simulation results showed that the response of the windshield greatly differed in terms of impact angles. At a less steep angle, the bird slid along the surface of the windshield and produced less damage to the structure. A steeper angle produced large deformation, and a plastic dimple could be observed at the impact point. These critical factors can be parameterized together to predict the combined effect and provide guiding principles for windshield design and optimization.

The proposed simulation methodology is helpful to analyze aeronautical and aerospace products under different impact scenarios and provides a comprehensive understanding of the impact event. Furthermore, the simulation methodology can be combined with various design approaches to optimize structural responses, and therefore it provides an alternative to costly and time-consuming experiments.

3.4. References

Abrate, S. (2016). Soft impacts on aerospace structures. *Progress in Aerospace Sciences*, 81, 1–17.

Ball, A. and McKenzie, H.W. (1994). On the low velocity impact behaviour of glass plates. *Journal de Physique IV Colloque*, 4, 783–788.

Barber, J.P., Taylor, H.R. and Wilbeck, J.S. (1978). Bird impact forces and pressures on rigid and compliant targets. No. UDRI-TR-77-17. University of Dayton Research Institute.

Belytschko, T., Rabczuk, T., Huerta, A. and Fernández-Méndez, S. (2004). Mesh free methods. *Encyclopedia of Computational Mechanics*. John Wiley & Sons.

Boehman, L.I. and Challita, A. (1982). A model for prediction bird and ice impact loads on structures. No. UDR-TR-79-54. GE Aviation, Ohio, USA.

Challita, A. (1980). Validation of a bird substitute for development and qualification of aircraft transparenciles. No. UDR-TR-80-69. University of Dayton Research Institute.

Chen, X.W. and Li, Q.M. (2004). Transition from non-deformable projectile penetration to semi-hydrodynamic penetration. *Journal of Engineering Mechanics*, 130, 123–127.

Cheng, C., Hiltner, A., Baer, E., Soskey, P. and Mylonakis, S. (1994). Deformation of rubber toughened polycarbonate: Macroscale analysis of the damage zone. *Journal of Applied Polymer Science*, 52(2), 177–193.

Cleary, E.C., Dolbeer, R.A. and Wright, S.E. (2006). Wildlife Strikes to Civil Aircraft in the United States 1990–2005. Federal Aviation Administration.

Crisfield, M.A., Remmers, J.J. and Verhoosel, C.V. (2012). *Nonlinear Finite Element Analysis of Solids and Structures*. John Wiley & Sons.

Dorogoy, A. and Rittel, D. (2015). Effect of confinement on thick polycarbonate plates impacted by long and AP projectiles. *International Journal of Impact Engineering*, 76, 38–48.

Dorogoy, A., Rittel, D. and Brill, A. (2011). Experimentation and modeling of inclined ballistic impact in thick polycarbonate plates. *International Journal of Impact Engineering*, 38(10), 804–814.

Doubrava, R. and Strnad, V. (2010). Bird strike analyses on the parts of aircraft structure. *27th Congress of the International Council of the Aeronautical Sciences*, Nice, France.

Duan, Y., Saigal, A., Greif, R. and Zimmerman, M.A. (2003). Impact behavior and modeling of engineering polymers. *Polymer Engineering & Science*, 43(1), 112–124.

Fraser, R.A.W. and Ward, I.M. (1977). The impact fracture behaviour of notched specimens of polycarbonate. *Journal of Materials Science*, 12(3), 459–468.

Grimaldi, A., Sollo, A., Guida, M. and Marulo, F. (2013). Parametric study of a SPH high velocity impact analysis: A birdstrike windshield application. *Composite Structures*, 96, 616–630.

Gunnarsson, C., Ziemski, B., Weerasooriya, T. and Moy, P. (2009). Deformation and failure of polycarbonate during impact as a function of thickness. *Proceedings of the SEM Annual Conference*, New Mexico, USA.

Hallquist, J.O. (2007). LS-DYNA Keyword User's Manual. Livermore Software Technology Corporation, 970.

Harewood, F.J. and McHugh, P.E. (2007). Comparison of the implicit and explicit finite element methods using crystal plasticity. *Computational Materials Science*, 39(2), 481–494.

Hu, W.K., Wang, Y.N., Yu, J., Yen, C.F. and Bobaru, F. (2013). Impact damage on a thin glass plate with a thin polycarbonate backing. *International Journal of Impact Engineering*, 62, 152–165.

Inberg, J.P.F. and Gaymens, R.J. (2002). Polycarbonate and co-continuous polycarbonate/abs blends: Influence of notch radius. *Polymer*, 40(15), 4197–4205.

Johnson, A.F. and Holzapfel, M. (2003). Modelling soft body impact on composite structures. *Composite Structures*, 61, 103–113.

Li, K. and Goldsmith, W. (1997). Perforation of steel and polycarbonate plates by tumbling projectiles. *International Journal of Solids and Structures*, 34(35–36), 4581–4596.

Liu, J., Li, Y.L. and Gao, X.S. (2014). Bird strike on a flat plate: Experiments and numerical simulations. *International Journal of Impact Engineering*, 70, 21–37.

Lombardo, B., Keskkula, H. and Paul, D. (1994). Influence of ABS type on morphology and mechanical properties of PC/ABS blends. *Journal of Applied Polymer Science*, 54(11), 1697–1720.

McCarthy, M.A., Xiao, J.R. and Mccarthy, C.T. (2004). Modelling of bird strike on an aircraft wing leading edge made from fibremetal laminates – Part 2: Modelling of impact with SPH bird model. *Applied Composite Materials*, 11, 317–340.

Qasim, H.S. (2009). Impact resistance of a rectangular polycarbonate armor plate subjected to single and multiple impacts. *International Journal of Impact Engineering*, 36(9), 1128–1135.

Qasim, H.S. and Yousif, A.A. (2008). Effect of distance from the support on the penetration mechanism of clamped circular polycarbonate armor plates. *International Journal of Impact Engineering*, 35(11), 1244–1250.

Sun, J.S., Lee, K.H. and Lee, H.P. (2000). Comparison of implicit and explicit finite element methods for dynamic problems. *Journal of Materials Processing Technology*, 105(1–2), 110–118.

Thorpe, J. (1997). The implications of recent serious bird strike accidents and multiple engine ingestions. *International Workshop on New Technologies for Bird Strike Prevention*, Boston, USA.

Tvergaard, V. and Needleman, A. (2008). An analysis of thickness effects in the Izod test. *International Journal of Solids and Structures*, 45(14–15), 3951–3966.

Wang, F.S. and Yue, Z.F. (2010). Numerical simulation of damage and failure in aircraft windshield structure against bird strike. *Materials & Design*, 31, 687–695.

Wang, F.S., Yue, Z.F. and Liu, Y.S. (2007). An optimization method to identify bird's constitutive model and its parameters. *Multidiscipline Model Materials and Structures*, 3, 527–534.

Wilbeck, J.S. (1978). Impact behavior of low strength projectiles. No. AFML-TR-77-134. Air Force Materials Lab Wright-Patterson AFB OH.

Wright, S.C., Fleck, N.A. and Stronge, W.J. (1993). Ballistic impact of polycarbonate – an experimental investigation. *International Journal of Impact Engineering*, 13(1), 120.

Yankelevsky, D. (1997). Local response of concrete slabs to low velocity missile impact. *International Journal of Impact Engineering*, 19(4), 331–343.

Zhu, S., Tong, M. and Wang, Y. (2009). Experiment and numerical simulation of a full-scale aircraft windshield subjected to bird impact. *50th AIAA/ ASME/ASCE/AHS/ASC Structures, Structural Dynamics, and Materials Conference*, Palm Springs, California, USA.

4

Integrated Simulation of Injection Molding Process and Mechanical Behavior

In the injection molding process, the melted polycarbonate is injected into a mold cavity of the desired shape and then cooled down under a high packing pressure. This process forces the polycarbonate to undergo a complex deformation and thermal history that affects the final properties of the product. Govaert *et al.* (Govaert *et al.* 2005; Engels *et al.* 2008, 2009a, 2009b) proved that the thermal history experienced from the melt to solidification has an impact on the yield stress of molded polycarbonate products. However, processing simulation tools such as Moldflow and Moldex3D only give access to the evaluation of shrinkage, warpage, etc., while the mechanical behavior of the final product is unknown. Likewise, mechanical analysis tools such as Abaqus and LS-DYNA can only evaluate the mechanical responses of products under desired loading conditions, while the effects of processing need to be further investigated.

The aim of this chapter is to present an integrated simulation framework. Based on the method of Govaert *et al.* (Govaert *et al.* 2005; Engels *et al.* 2008, 2009a, 2009b), the inhomogeneous yield stress of polycarbonate is first evaluated and experimentally verified. It is then incorporated into the elastic–viscoplastic constitutive model of polycarbonate presented in Chapter 2 and further applied to simulate the mechanical behavior of polycarbonate.

To verify the integrated simulation methodology, the notched Izod impact tests are carried out to analyze polycarbonate specimens undergoing different thermal histories. The integrated simulation is performed to reveal the influence of thermal history on the Izod impact behavior of polycarbonate. Comparisons and discussions about the results of fracture energy obtained numerically and experimentally are undertaken. It is shown that a satisfied agreement exists between the two.

4.1. Yield stress modeling from thermal history

4.1.1. *Govaert's yield stress model*

In the injection molding process, the thermal history from melting to solidification is a cooling down process. According to Govaert *et al.* (Govaert *et al.* 2005; Engels *et al.* 2008, 2009a, 2009b), the physical evolution of yield stress during this process is equivalent to the evolution of yield stress during the annealing process. In this context, the tensile test specimens of polycarbonate (Lexan 141R) are first realized by the injection molding machine LIGUANG FL-220MO at two different mold temperatures (90 and 140°C). The geometry of a molded dog-bone specimen is shown in Figure 4.1.

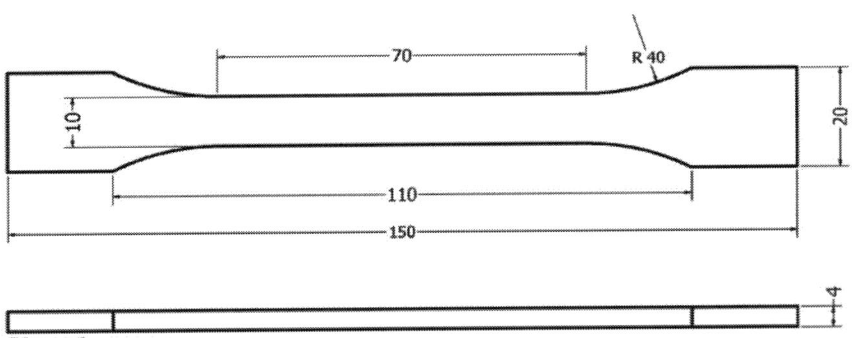

Figure 4.1. *Geometry of a tensile test specimen*

The glass transition temperature (T_g) of polycarbonate is determined to be 148°C using the dynamic mechanical thermal analysis (DMTA) (Xu *et al.* 2015; Dar *et al.* 2017). At high temperatures, especially for the range near T_g, the increase in yield stress is more pronounced, as observed by Golden *et al.* (1967) and Bauwens-Crowet and Bauwens (1982). In this study, all the annealed test specimens are kept for a specified period of time at four different annealing temperatures, T_a (80, 100, 120 and 130°C), and then slowly cooled down to room temperature. Thereafter, tensile tests are performed under a constant strain rate of 10^{-2} s^{-1} to determine the yield stress of all the test specimens on a CRIMS DNS 100 electronic universal testing system.

Figure 4.2 shows the evolution of yield stress of test specimens obtained at the mold temperature of 90°C and four different annealing temperatures. It is observed that the yield stress increases with the annealing time, and one such increase is more pronounced at high temperatures, which coincides well with the existing observations (Engels *et al.* 2009a; Golden *et al.* 1967; Bauwens-Crowet and Bauwens 1982). Note that the increase in yield stress cannot be endless. When the thermodynamic equilibrium of polycarbonate is reached after a long annealing time, the yield stress stops growing and reaches a plateau value (Bauwens-Crowet and Bauwens 1982).

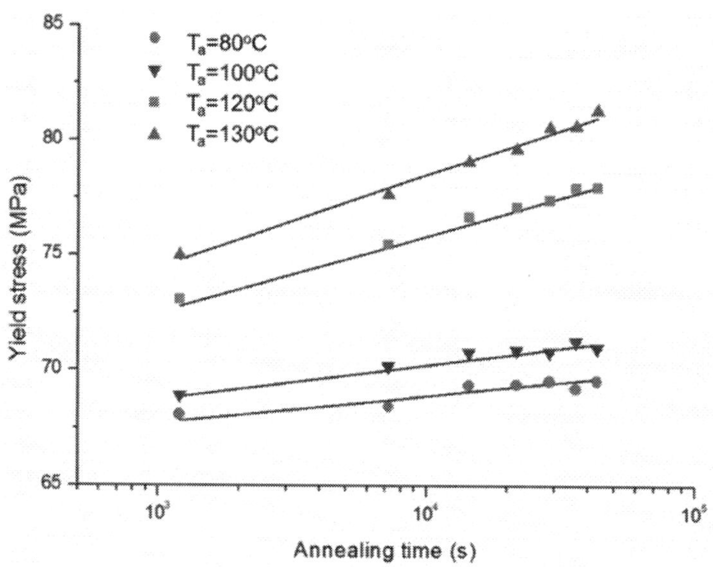

Figure 4.2. *Evolution of yield stress of test specimens obtained at the mold temperature of 90°C and four different annealing temperatures. For a color version of this figure, see www.iste.co.uk/zhang/polycarbonate.zip*

The evolution curves of the yield stress at various temperatures can be combined into a single master curve for a reference temperature using the time–temperature superposition (TTS) (Ronan *et al.* 2007; Zhao *et al.* 2008). It is generally recognized that the mechanical properties of a polymer are influenced by both the temperature and the temperature loading time. The mechanical properties tested over a short time and in a high-temperature condition are equivalent to the mechanical properties tested over a long time and in a low-temperature condition (Zhao *et al.* 2008). This equivalence can be quantitatively described by the TTS, considered to be one of the most important principles in polymer physics.

In detail, because the yield stress function $\sigma_y(t)$ at one temperature is similar to the same functions at neighboring temperatures (Verbeeten *et al.* 2015), the curves of yield stress versus logarithmic time at different temperatures can horizontally be shifted along the axis of logarithmic time to superpose a master curve. The shift distance along the axis of logarithmic time is known as the time–temperature shift factor $a_T(T)$, which can be described by the Arrhenius relation

$$a_T(T) = \exp\left(\frac{\Delta U_a}{R} \cdot \left(\frac{1}{T} - \frac{1}{T_{ref}}\right)\right) \qquad [4.1]$$

where ΔU_a is the activation energy equal to 205 kJ/mol (Govaert *et al.* 2005; Engels *et al.* 2008), *R* is the universal gas constant and T_{ref} is the reference temperature.

In Figure 4.3, the master curves are presented for the yield stress of specimens at a reference temperature of 23°C (room temperature) and mold temperatures of 90 and 140°C. It can be observed that: (1) two "ineffective" periods of time denoted as t_a^1 and t_a^2 exist during which the yield stress remains nearly unchanged. However, both the "ineffective" period of time and the initial value of yield stress vary with the mold temperature. This is because the processed polycarbonate specimen has an initial thermodynamic state that is related to the thermal processing history. The parameter t_a represents the time that a specimen without the thermal processing history needs to be annealed at room temperature to reach the same thermodynamic state. As a result, if the annealing time is less than t_a, the yield stress remains unchanged. (2) After a certain annealing time, both master curves coincide. With the increase in annealing time, the thermodynamic equilibrium of polycarbonate is reached and the yield stress eventually reaches a constant value.

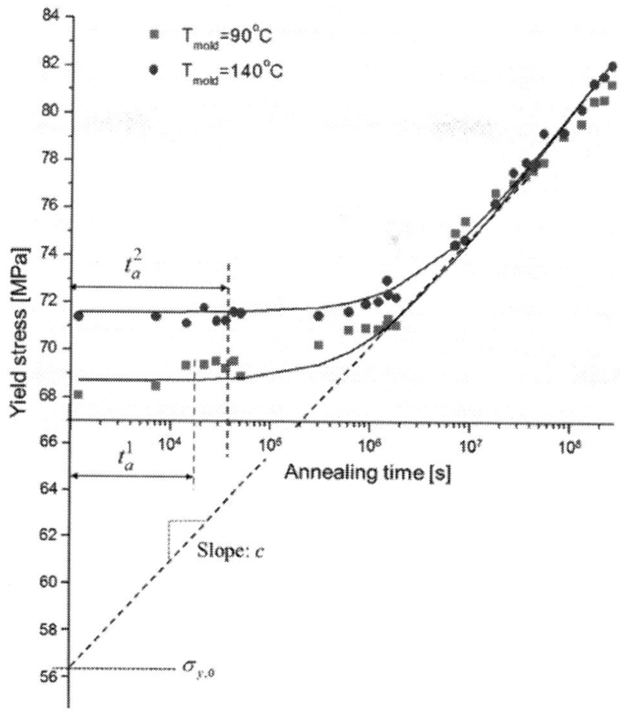

Figure 4.3. *Master curves for the evolution of yield stress of polycarbonate test specimens at two different mold temperatures. For a color version of this figure, see www.iste.co.uk/zhang/polycarbonate.zip*

Correspondingly, a logarithmic evolution equation of the yield stress was proposed by Govaert *et al.* (Govaert *et al.* 2005; Engels *et al.* 2008; Engels *et al.* 2009b) to model the annealing effect:

$$\sigma_y(t) = \sigma_{y,0} + c \cdot \log\left(t_{eff}^c + t\right) \quad [4.2]$$

where $\sigma_{y,0}$ is the intersection of the tangent of the master curve with the yield stress axis, c is the slope and t_{eff}^c is the effective accumulated time. It only depends on the thermal processing history, which is defined in terms of $T_c(t)$:

$$t_{eff}^c = \int_0^{t_c} a_T^{-1}\left(T_c(t)\right)dt \quad [4.3]$$

where $T_c(t)$ is the thermal processing history during cooling down. The initial value of yield stress shown in Figure 4.3 can be determined at $t_a = 0$ so that equation [4.2] becomes

$$\sigma_y = \sigma_{y,0} + c \cdot \log\left(\int_0^{t_c} a_T^{-1}\left(T_c\left(t\right)\right)dt\right) \qquad [4.4]$$

At a strain rate of 10^{-2} s^{-1}, the values of $\sigma_{y,0}$ and c are directly obtained from the master curves as $\sigma_{y,0} = 56.4\,\text{MPa}$ and $c = 3.8$ MPa.

Here, the thermal history, $T_c(t)$, is determined by means of Moldflow (Shoemaker 2006). The rheological properties of Lexan 141R are taken from the standard Moldflow material library.

4.1.2. *Experimental validations of the yield stress model*

As an example, Figure 4.4 shows the thermal histories and thermal maps at different instances over the thickness in the center of the dog-bone specimen. Here, the mold temperature is 40°C. The melt temperature and the injection velocity are set to 280°C and 50 cm^3/s respectively. The packing pressure and the cooling time are set to 50 MPa and 70 s respectively. It can be seen that the temperature is always higher at the center than near the surface. The cooling rate is maximum at the surface and minimum in the center. The thermal history of each element is then exported into equation [4.4] to obtain the initial yield stress. Obviously, a heterogeneous yield stress distribution is obtained over the specimen.

To assess the yield stress model, tensile test specimens are injection-molded at the mold temperature ranging from 40 to 130°C in steps of 10°C. Likewise, the melt temperature and the injection velocity are set to 280°C and 50 cm^3/s respectively. The packing pressure and the cooling time are set to 50 MPa and 70 s respectively. The yield stress at different mold temperatures is subsequently determined in uniaxial tension with a strain rate of 10^{-2} s^{-1} at room temperature.

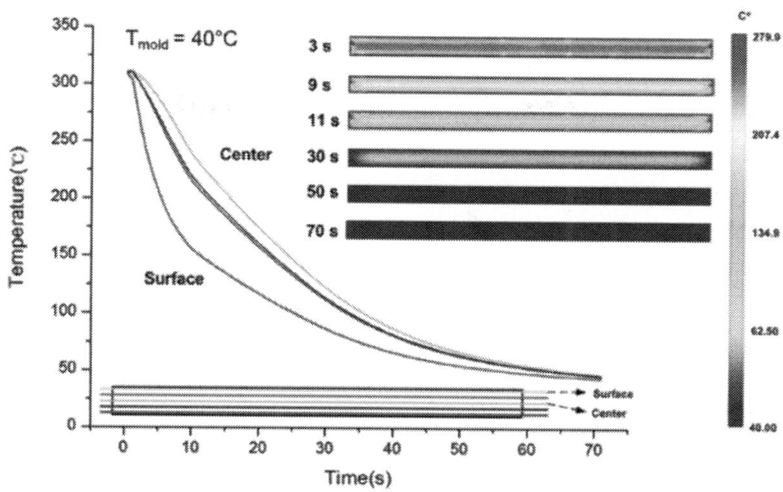

Figure 4.4. *Thermal histories $T_c(t)$ at different locations. For a color version of this figure, see www.iste.co.uk/zhang/polycarbonate.zip*

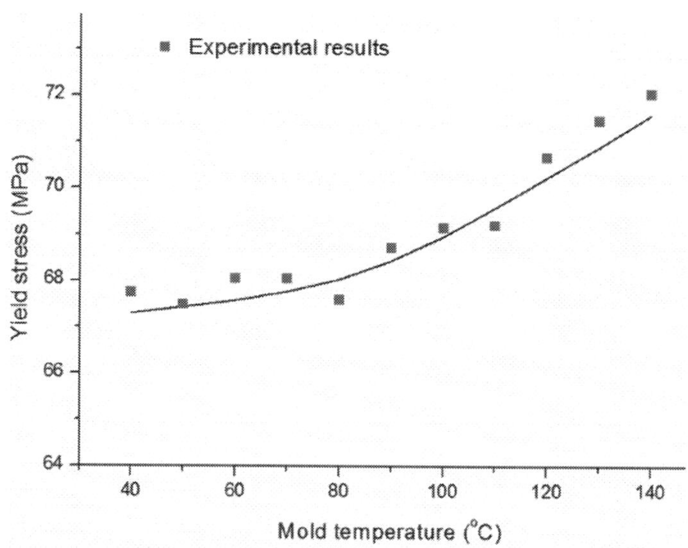

Figure 4.5. *Yield stress versus mold temperature. For a color version of this figure, see www.iste.co.uk/zhang/polycarbonate.zip*

A thickness-weighted average (Govaert *et al.* 2005; Engels *et al.* 2008; Engels *et al.* 2009b) of the computed yield stress is used for comparison with the experiments. Figure 4.5 compares the average yield stress with the experimental results. Clearly, a good agreement exists and the evolution of the yield stress versus mold temperature is well demonstrated.

4.2. Setup of the Izod impact test

In accordance with to the ISO 180-2000 standard (ISO 180 2000), notched Izod impact tests are carried out. The impact behavior of polycarbonate specimens is analyzed. Rectangular plates of polycarbonate Lexan 141R, with dimensions 170 × 70 × 3.2 mm^3, are injection-molded. The melt temperature and the injection velocity are set to 295°C and 30 cm^3/s respectively. The packing pressure and the cooling time are set to 65 MPa and 60 s respectively. Three different mold temperatures 60, 80 and 120°C are used for injection molding. Rectangular bars of dimensions 80 × 10 × 3.2 mm^3 are cut from the polycarbonate plates. The V-shaped notch with 3 mm in thickness and 30° is generated in the center of the bars by a slot milling machine. The geometry of a notched Izod impact specimen is shown in Figure 4.6.

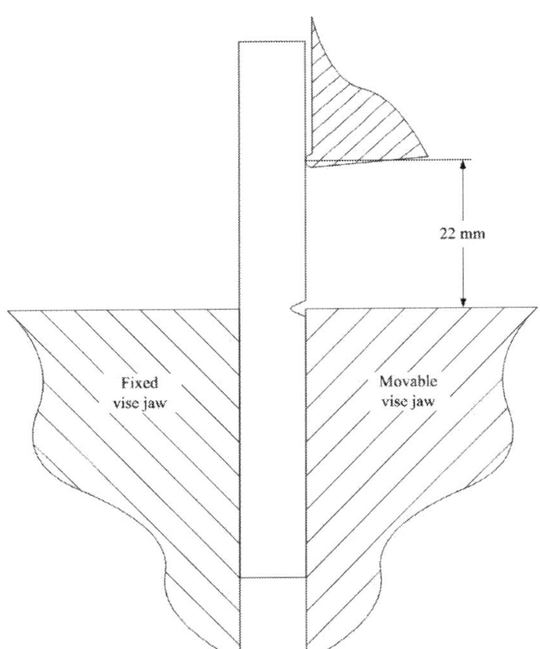

Figure 4.6. *Illustration of the notched Izod impact test*

The impact test is conducted with a pendulum-type impactor CEAST 9050. During the impact test, the bottom of the specimen is clamped as a cantilever beam. The upper part of the specimen is hit by the striking nose of the pendulum hammer with a controlled energy. As shown in Figure 4.6, the distance between the fixed support and the contact position of the hammer's striking nose is kept at 22 mm according to the standard. A calibrated hammer with a mass of 0.6 kg is used in the present test. The magnitudes of the initial impact energy and velocity are controlled by changing the initial angle of the hammer to fully fracture the polycarbonate specimens. Impact tests are performed to determine the fracture energies of the specimens realized at different mold temperatures (60, 80 and 120°C). The results are given in Table 4.1.

4.3. Integrated simulation framework

The integrated simulation framework is schematically shown in Figure 4.7. Moldflow and LS-DYNA are commercial finite element tools used to conduct the processing simulation and mechanical simulation respectively.

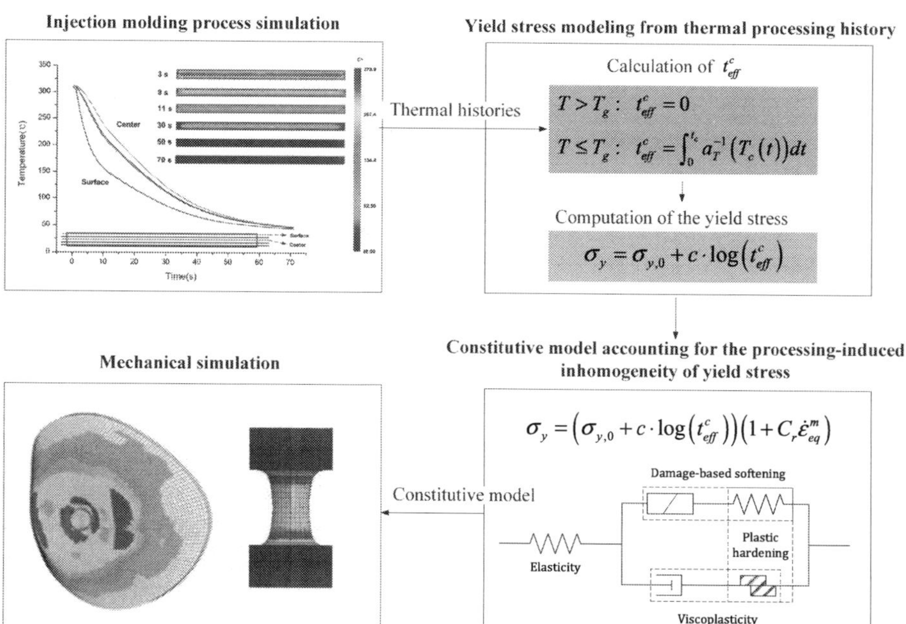

Figure 4.7. *Illustration of the integrated simulation framework. For a color version of this figure, see www.iste.co.uk/zhang/polycarbonate.zip*

– In the first step, the finite element model of the product, developed in LS-DYNA, is imported into Moldflow for simulation of the injection molding process. Note that the meshes of the product in the processing simulation model should be exactly the same as the meshes of the mechanical simulation model. In this case, the calculated yield stress from the thermal processing history can be directly adopted in the mechanical simulation without interpolations.

– In the second step, the thermal histories of all the elements during processing are obtained after the processing simulation, which are further exported to equation [4.4] to calculate the yield stress. Obviously, different thermal histories of the elements lead to the heterogeneous distribution of yield stress in the product.

– In the third step, the elastic–viscoplastic constitutive model of polycarbonate is modified to incorporate the processing-induced inhomogeneity of yield stress.

– Finally, mechanical simulations are performed in LS-DYNA. The modified constitutive model is applied to model the properties of polycarbonate. In this way, the influence of processing is taken into account in the mechanical simulations.

The integrated simulation is then performed to model the Izod impact behavior of polycarbonate and compared with experimental results.

4.4. Integrated simulation of the injection molding process and Izod impact

4.4.1. *Prediction of yield stress based on the injection molding simulation*

A 3D finite element model of the specimen used for the injection molding simulation is shown in Figure 4.8. A total of 36,368 tetrahedral elements are used to model the specimen. Since the deformation is mostly concentrated in the vicinity of the notch, a refined mesh is generated for this region. The process variables used for the injection molding simulation are identical to those in the experiment. The thermal history of each element is obtained at three different mold temperatures (60, 80 and 120°C), which is then exported into equation [4.4] to calculate the yield stress.

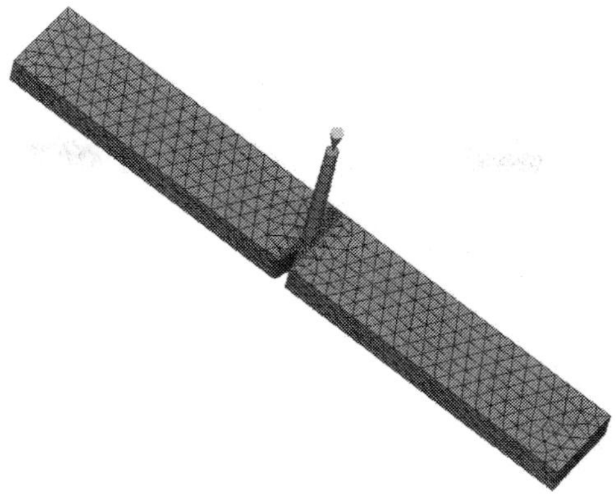

Figure 4.8. *Finite element model of the specimen*

For a mold temperature of 60°C, the cooling profiles along the central cross-section of the specimen are calculated, as shown in Figure 4.9. As expected, the cooling is fast near the surface and slow at the center of the specimen.

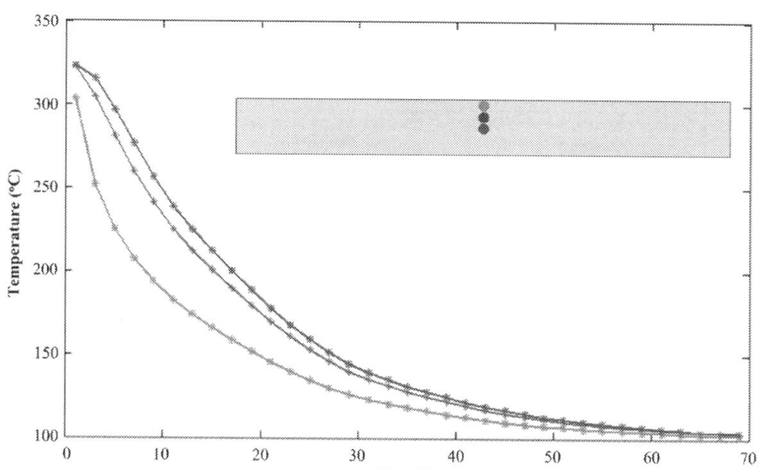

Figure 4.9. *Cooling profiles along the central cross-section of the specimen. For a color version of this figure, see www.iste.co.uk/zhang/polycarbonate.zip*

The distributions of yield stress within the central cross-section are shown in Figure 4.10. A much lower yield stress can be found near the surface than in the center.

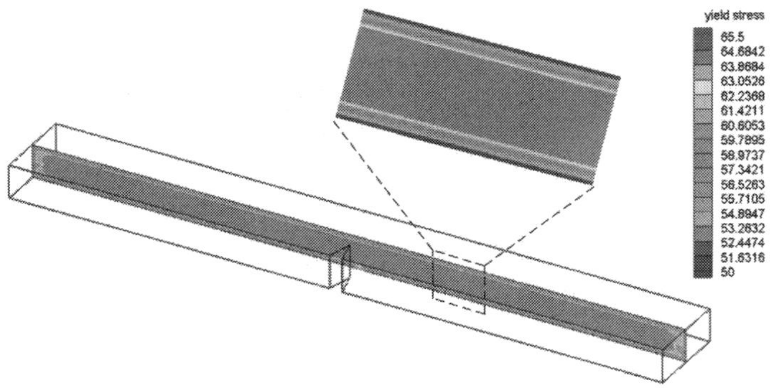

Figure 4.10. *Distributions of yield stress within the central cross-section. For a color version of this figure, see www.iste.co.uk/zhang/polycarbonate.zip*

Figure 4.11 shows the distributions of yield stress within the central cross-section taken from the specimen at different mold temperatures. The yield stress distributions in the specimens at different mold temperatures are coincident. However, the value of yield stress is increased as the mold temperature increases.

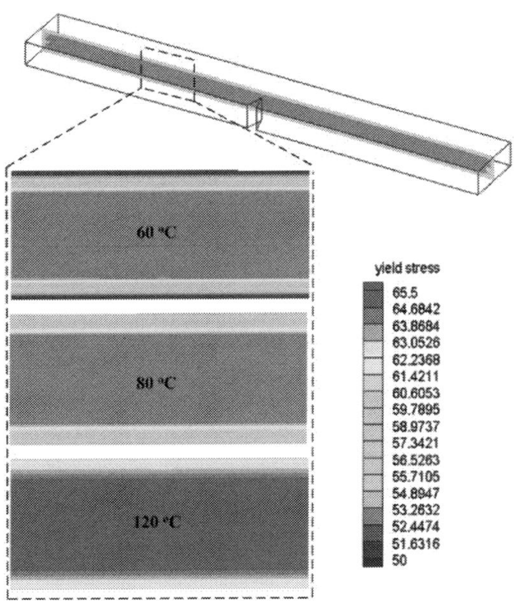

Figure 4.11. *Yield stress distributions of the specimens at different mold temperatures. For a color version of this figure, see www.iste.co.uk/zhang/polycarbonate.zip*

4.4.2. *Constitutive model with the processing-induced inhomogeneity of yield stress*

In Chapter 2, an elastic–viscoplastic constitutive model is presented to characterize the mechanical properties of polycarbonate over a wide range of strain rates and temperatures. The yield stress of polycarbonate is formulated as a function depending on the strain rate and temperature:

$$\sigma_y = K\left(1 + C_r \dot{\varepsilon}_{eq}^m\right)\left(1 + C_t T_h\right) \qquad [4.5]$$

where C_r and m are strain rate-dependent parameters, K is the yield stress tested at the reference strain rate and temperature, C_t is a temperature-dependent parameter, $\dot{\varepsilon}_{eq}^m$ is the equivalent strain rate and T_h is the homologous temperature.

In brief, the effect of processing-affected yield stress related to equation [4.4] can be introduced into equation [4.5]. The following features should be noted.

First, only the impact problems at room temperature are taken into account at present. Moreover, although a high strain rate deformation may result in the rise of local temperature in polymers, this influence is found to be modest (Sarva and Boyce 2007; Tvergaard and Needleman 2008). Therefore, the temperature-dependent terms in equation [4.5] are neglected. The new formulation for yield stress is expressed as:

$$\sigma_y = \left(\sigma_{y,0} + c \cdot \log\left(t_{eff}^c\right)\right)\left(1 + C_r \dot{\varepsilon}_{eq}^m\right) \qquad [4.6]$$

Second, the strain rate sensitivity parameter C_r and m are identified according to the tensile test results of the specimens at strain rates ranging from 10^{-4} to 10^3 s^{-1}. The identified values are: $m = 2.815$ and $C_r = 0.079$.

Material properties such as elastic modulus E, Poisson's ratio ν, density ρ as well as the other parameters of the constitutive model are kept the same (see Table 2.2 in Chapter 2). The modified constitutive model is then implemented into LS-DYNA by using a user-defined material subroutine (UMAT). Details of the implemented numerical algorithm can be found in Chapter 2.

4.4.3. *Impact simulation*

3D finite element models of the specimen and pendulum hammer are shown in Figure 4.12. The mesh of the specimen in the impact simulation is exactly the same as in the processing simulation model. Note that the hummer is modeled as a reduced part to save the computing time; however, it has the equivalent mass of the hummer in terms of the kinematic energy (Silberstein 2005).

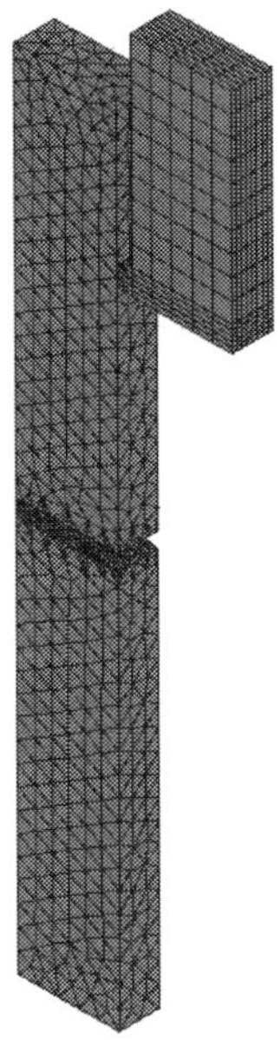

Figure 4.12. *Finite element models of the specimen and the pendulum hammer*

The bottom of the specimen is constrained to avoid any translational and rotational motion, which exercises a fully clamped boundary condition. The impactor is modeled as a rigid body that is allowed to move along the direction normal to the impact surface. The initial horizontal velocity of the hummer is 3.5 m/s. For an impact problem, it has been shown that the rigid assumption of the impactor effectively reduces the computing time and has minor effects on the numerical results (Ullah *et al.* 2013; Louche *et al.* 2009). Therefore, the hammer is assigned with Young's modulus of 200 GPa and Poisson's ratio of 0.3. Meanwhile, all contact surfaces are assumed to be frictionless. The contact model **CONTACT_ERODING_NODES_TO_SURFACE* in LS-DYNA is used to simulate the contact boundaries. The non-interpenetration condition between the hammer and the specimen is ensured by using the penalty-based contact algorithm (DYNA 2006).

4.4.4. *Simulation results and comparisons with experiments*

In this section, numerical results for the notched Izod impact of polycarbonate specimens at different mold temperatures are presented and compared with experiments.

The numerical results of the plastic strain distribution within the undamaged specimens at different mold temperatures are presented in Figure 4.13. The greatest straining occurs at the notch, and there is also a high-strain region opposite to the notch where the bending strains are concentrated. The plastic strain is always higher for the specimen with lower mold temperature than with the opposite. This is due to the processing-affected yield stress of polycarbonate. Since the yield stress of polycarbonate is reduced as the mold temperature decreases, plastic deformation will be more easily developed and the plastic strain will be accumulated to a higher level in the specimen with lower mold temperature.

Figure 4.14 presents the variation of internal energy, i.e. the total energy absorbed by the specimens at different mold temperatures during the impact. Internal energies can be found to vary in a similar way at different mold temperatures. The energies gradually increase with time and suddenly decrease at the final stage, which means that the specimens are fractured. Since the elastic properties of polycarbonate specimens at different mold temperatures are assumed to be constant, the variations of the energy with time are exactly the same before the yielding of polycarbonate. Thus, the energy variation curves of the specimens at different mold temperatures are overlapping by up to nearly 3 ms. However, distinct separations of the energy curves are found after yielding, especially in the fracture process.

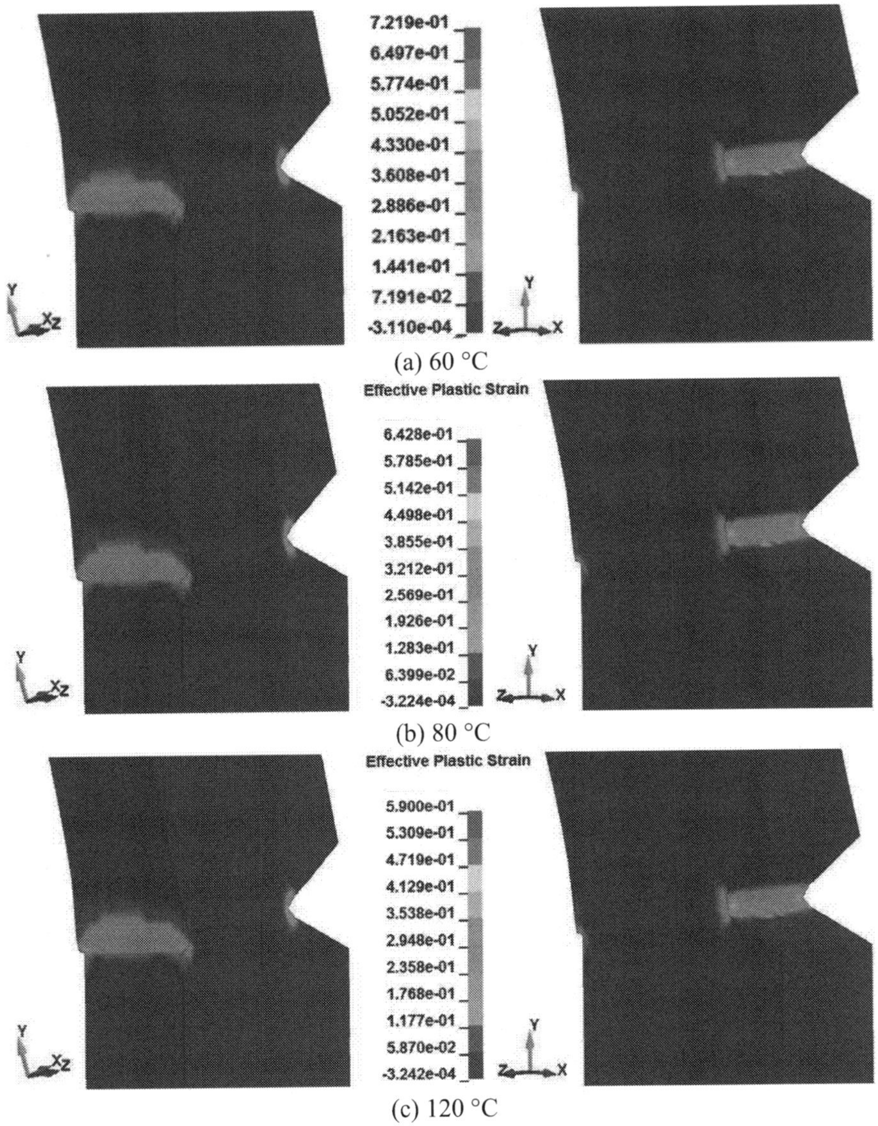

Figure 4.13. *Plastic strain distribution within the undamaged specimens at different mold temperatures. For a color version of this figure, see www.iste.co.uk/zhang/polycarbonate.zip*

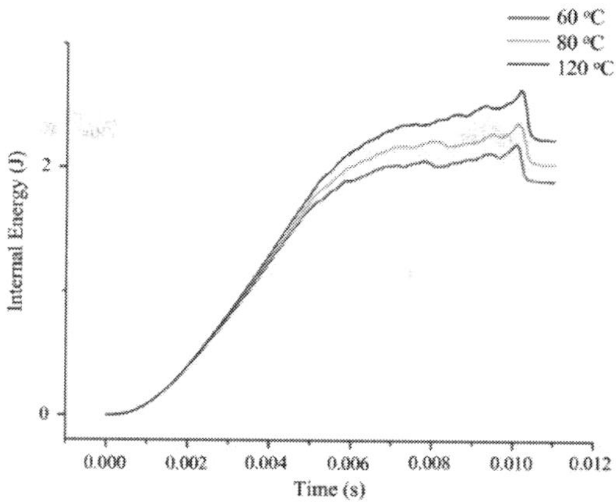

Figure 4.14. *Internal energy variation of impacted specimens at different mold temperatures. For a color version of this figure, see www.iste.co.uk/zhang/polycarbonate.zip*

Figure 4.15 shows the detailed fracture process of the specimen at the mold temperature of 60°C.

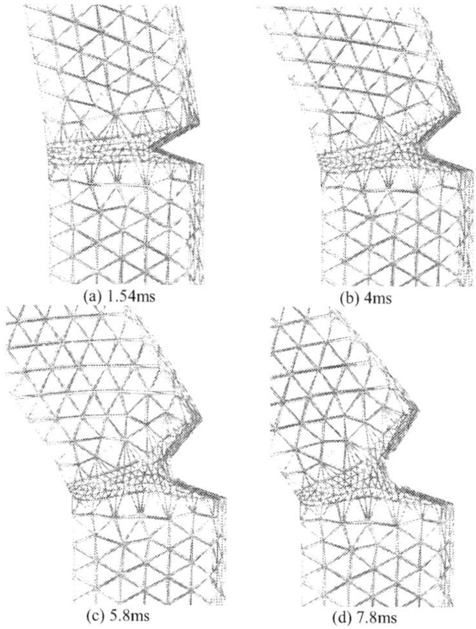

Figure 4.15. *Fracture process of the specimen at the mold temperature of 60°C. For a color version of this figure, see www.iste.co.uk/zhang/polycarbonate.zip*

Table 4.1 lists the fracture energies obtained by numerical simulations and experimental tests. The fracture energy is measured by dividing the total lost energy by the cross-sectional area at the notch (ISO 180 2000).

Mold temperature (°C)	60	80	120
Lost energy of the striker (J)	2.1708	2.3067	2.4343
Numerical fracture energy (kJ/m^2)	77.53	82.38	86.94
Experimental fracture energy (kJ/m^2)	80.90	85.40	90.10

Table 4.1. *Fracture energies of numerical simulations and experimental tests*

As shown in Table 4.1, both experimental and numerical results clearly indicate that the fracture energy increases as the mold temperature increases. Since the increase in the mold temperature leads to the increase in the yield stress of polycarbonate, more energy of the striker will be consumed for the plastic deformation and fracture of the polycarbonate specimen. The predicted fracture energies of the polycarbonate specimens at 60, 80 and 120°C mold temperatures are shown to be in good agreement with the experimental data, indicating the predictive ability of the proposed integrated simulation framework.

4.5. Conclusion

In this chapter, an integrated simulation framework was developed to incorporate the inhomogeneous distribution of yield stress induced by injection molding into the mechanical simulations of polycarbonate products. A modeling approach developed by Govaert *et al.* (Govaert *et al.* 2005; Engels *et al.* 2008; Engels *et al.* 2009b), which allows the prediction of the yield stress from the thermal history during the injection molding process, was used in this study. An experimental test and finite element simulation of the notched Izod impact test were performed to analyze the impact behavior of polycarbonate specimens at different mold temperatures. A constitutive model with a modification to combine the influence of the thermal processing history was developed to model the mechanical properties of polycarbonate. The findings from the numerical and experimental results can be summarized as follows:

1) The fracture energies of the polycarbonate specimens at different mold temperatures obtained by the numerical simulation coincide well with the

experimental test, indicating the predictive ability of the proposed framework.

2) Since the yield stress of polycarbonate is significantly influenced by the model temperature, the plastic deformation and fracture energies of polycarbonate specimens display obvious differences at various mold temperatures. Both the experimental and numerical results indicate that the plastic strain and fracture energy increase with the increase in the mold temperature.

4.6. References

Bauwens-Crowet, C. and Bauwens, J.C. (1982). Annealing of polycarbonate below the lass transition: Quantitative interpretation of the effect on yield stress and differential scanning calorimetry measurements. *Polymer*, 23(11), 1599–1604.

Dar, U.A., Xu, Y.J., Zakir, S.M. and Saeed, M.U. (2017). The effect of injection molding process parameters on mechanical and fracture behavior of polycarbonate polymer. *Journal of Applied Polymer Science*, 134(7), 44474.

DYNA (2006). LS-DYNA Theory Manual. Livermore Software Technology Corporation.

Engels, T.A.P., Govaert, L.E. and Meijer, H.E.H. (2008). Quantitative prediction of mechanical performance of polymer products directly from processing conditions. *American Institute of Physics Conference Proceedings*, 1027, 1336.

Engels, T.A.P., Govaert, L.E. and Meijer, H.E.H. (2009a). The Influence of molecular orientation on the yield and post-yield response of injection-molded polycarbonate. *Macromolecular Materials and Engineering*, 294(12), 821–828.

Engels, T.A.P., van Breemen, L.C.A., Govaert, L.E. and Meijer, H.E.H. (2009b). Predicting the long-term mechanical performance of polycarbonate from thermal history during injection molding. *Macromolecular Materials and Engineering*, 294(12), 829–838.

Golden, J.H., Hammant, B.L. and Hazell, E.A. (1967). Effect of thermal pretreatment on the strenght of polycarbonate. *Journal of Applied Polymer Science*, 11(8), 1571–1579.

Govaert, L.E., Engels, T.A.P., Klompen, E.T.J., Peters, G.W.M. and Meijer, H.E.H. (2005). Processing-induced properties in glassy polymers. *International Polymer Processing*, 20(2), 170–177.

International Organization of Standardization (2000). Plastics-Determination of Izod impact strength.

Louche, H., Piette-Coudol, F., Arrieux, R. and Issartel, J. (2009). An experimental and modeling study of the thermomechanical behavior of an ABS polymer structural component during an impact test. *International Journal of Impact Engineering*, 36, 847–861.

Ronan, S., Alshuth, T., Jerrams, S. and Murphy, N. (2007). Long-term stress relaxation prediction for elastomers using the time-temperature superposition method. *Material & Design*, 28, 1513–1523.

Sarva, S.S. and Boyce, M.C. (2007). Mechanics of polycarbonate during high-rate tension. *Journal of Mechanics of Materials and Structures*, 2, 1853–1880.

Shoemaker, J. (2006). *Moldflow Design Guide*. Carl Hanser Verlag, Munich.

Silberstein, M.N. (2005). Mechanics of notched Izod impact testing of polycarbonate. Master's Thesis, Massachusetts Institute of Technology.

Tvergaard, V. and Needleman, A. (2008). An analysis of thickness effects in the Izod test. *International Journal of Solids and Structures*, 45, 3951–3966.

Ullah, H., Harland, A.R. and Silberschmidt, V.V. (2013). Damage and fracture in carbon fabric reinforced composites under impact bending. *Composite Structures*, 101, 144–156.

Verbeeten, W.M.H., Kanters, M.J.W. and Engels, T.A.P. (2015). Yield stress distribution in injection-moulded glassy polymers. *Polymer International*, 64, 1527–1536.

Xu, Y.J., Lu, H., Gao, T.L. and Zhang, W.H. (2015). Predicting the low-velocity impacted behavior of polycarbonate: Influence of thermal history during injection molding. *International Journal of Impact Engineering*, 86, 265–273.

Zhao, J., Knauss, W.G. and Ravichandran, G. (2008). Applicability of the time-temperature superposition principle in modeling dynamic response of a polyuria. *Mechanics of Time-Dependent Materials*, 11, 289–308.

5

Process Optimization of the Injection Molding for High Mechanical Performance

Aeronautical and aerospace polycarbonate products with a curved and thin-walled shape are usually produced by the injection molding process. The dimensional accuracy and the mechanical performance of these products exhibit significant dependence on processing conditions. In the injection molding process, warpage caused by non-uniform shrinkage during the processing has been considered to be the most serious defect of thin-walled products (Courbebaisse and Garcia 2002; Gu *et al.* 2001). In fact, the dimensional accuracy of thin-walled products mainly relies on whether the processing parameters are reasonably selected to reduce or avoid the warpage of the molded product. Therefore, in recent years, considerable attempts have been made to optimize injection molding process parameters (Zhou *et al.* 2006; Shen *et al.* 2007; Kurtaran *et al.* 2005; Kurtaran and Erzurumlu 2006; Yin *et al.* 2011a; Yin *et al.* 2011b; Ozcelik and Erzurumlu 2006).

However, in aeronautical and aerospace applications, the ultimate aim of the injection molding process is not only to satisfy the dimensional accuracy, but also to acquire an excellent mechanical performance of the product. The influence of the process on the mechanical performance of injection-molded product arises from two aspects: first, as discussed in Chapter 4, the non-uniform thermal history during the injection molding process induces the inhomogeneity of yield stress and consequently affects the mechanical behavior of polycarbonate products. Second, due to the processing-induced warpage deformation, the molded product will be forced to deform from the "warped shape" to the "designed shape" in the assembling process. Thus, residual stress is generated within the injection-molded product, which has a remarkable influence on the mechanical performance of the product.

Therefore, if injection molding process parameters can be adjusted in an intelligent way, the mechanical performance of the product can be improved towards the expected service conditions. In this chapter, we present a process optimization methodology to improve the mechanical performance of polycarbonate products under impact loading. The relationship between mechanical performance and processing parameters is determined by incorporating the warpage and inhomogeneous yield stress into the simulation. An integrated simulation framework combining the process simulation and the mechanical simulation is developed. This integrated framework can be regarded as an extension of the integrated framework presented in Chapter 4 and is presented in Figure 5.1.

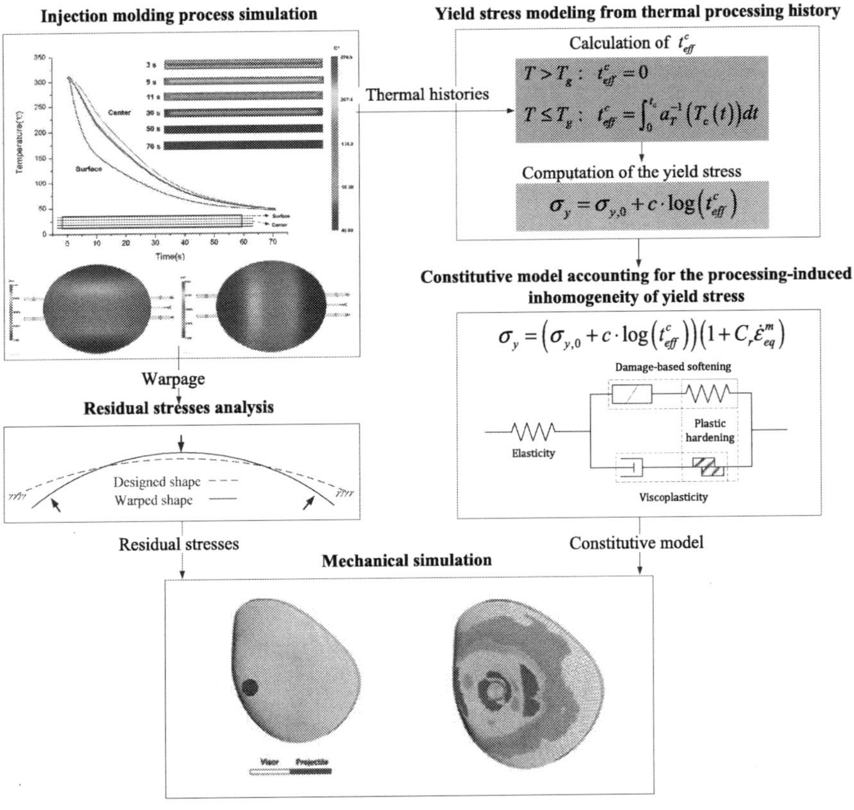

Figure 5.1. *Integrated simulation framework. For a color version of this figure, see www.iste.co.uk/zhang/polycarbonate.zip*

Undoubtedly, costly finite element computations are not suitable for a large number of repeated analyses required in an optimization process. Therefore, in the proposed optimization methodology, a prediction model created by the back propagation (BP) neural network (Fogel and Robinson 2003) is used to construct a mathematical approximation that maps the relationship between mechanical performances and processing parameters to replace expensive finite element analyses. The particle swarm optimization (PSO) algorithm (Kennedy and Eberhart 1995; Fourie and Groenwold 2002) is then interfaced with this prediction model to optimize processing parameters for the improvement of mechanical performance.

5.1. Integrated simulation framework of an astronaut's helmet visor

Here, an astronaut's helmet visor made of polycarbonate is studied. Typical processing parameters, including mold temperature, melt temperature, injection velocity, compression distance, compression force, compression velocity and compression waiting time, are considered as design variables. Values of processing parameters are optimized to minimize the maximum von Mises stress within the polycarbonate visor against the impact.

In Figure 5.1, the processing simulation software Moldflow is first used to simulate the injection molding process and to obtain the warpage and thermal histories. Here, it should be noted that the finite element model of the visor developed in LS-DYNA is imported into Moldflow for processing simulation. As the meshes of the processing simulation model are the same as the meshes of the impact simulation model, the calculated yield stress from the thermal processing history can be directly adopted in the impact simulation without interpolations.

Second, the warped visor is clamped and forced to deform to its expected shape. The generated residual stresses are calculated by using LS-DYNA. Meanwhile, the thermal histories obtained after the injection molding process simulation are exported to calculate the yield stresses. Then, the elastic–viscoplastic constitutive model of polycarbonate is modified to incorporate the processing-induced inhomogeneity of yield stress.

Finally, mechanical simulations are performed in LS-DYNA. The residual stresses are imposed on the finite element model of the visor as the prestress.

The modified constitutive model is applied to model the properties of polycarbonate. In this way, the influence of processing is taken into account in the mechanical simulations.

5.1.1. *Simulation of the injection molding process*

A geometrical model of the helmet visor used in this study is created according to the real structure shown in Chapter 3 and meshed in LS-DYNA. The finite element model is then imported into Moldflow for the injection molding process simulation. Figure 5.2 shows the finite element model of the visor with the injecting gate and the cooling channel. The visor is meshed with 43,200 3D solid elements. In order to keep the surface quality of the visor, the injecting gate uses the side gate form, as shown in Figure 5.2. The material used is polycarbonate Lexan 141R. The rheological properties of Lexan 141R are taken from the standard Moldflow material library.

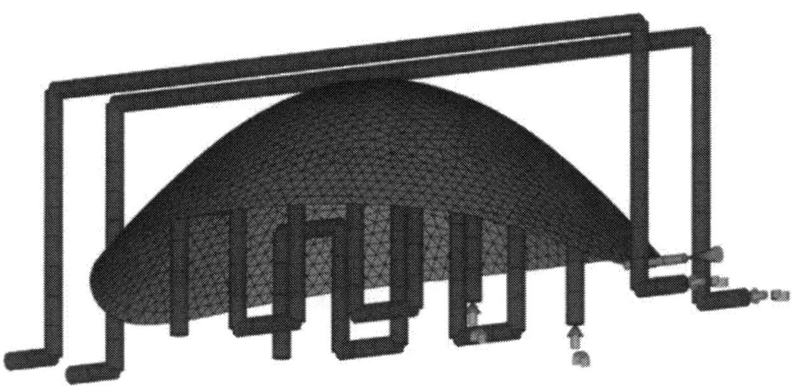

Figure 5.2. *Finite element model with the injecting gate and the cooling channel*

Finite element analysis is carried out using the recommended process parameters in Moldflow, as listed in Table 5.1. The computed warpage results in the x, y and z directions are shown in Figure 5.3. The coordinates of each node of the warped visor are exported and denoted as x_i, y_i and z_i (where $i = 1, 2, \ldots, n$, n is the node number). In addition, warpage values at all the nodes are also exported and denoted as W_i^x, W_i^y and W_i^z.

(a) *x* direction

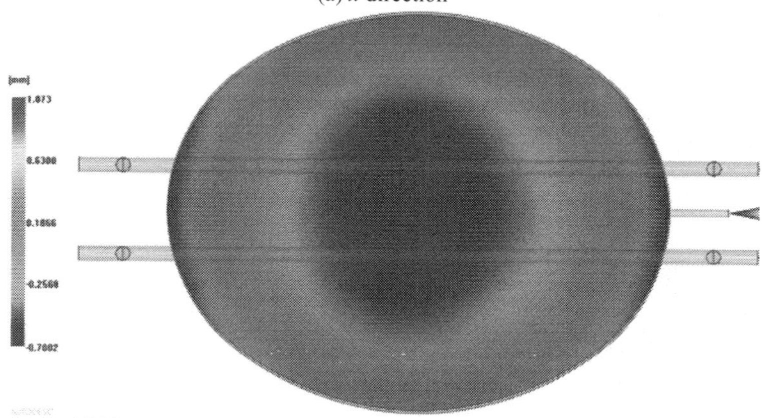

(b) *y* direction

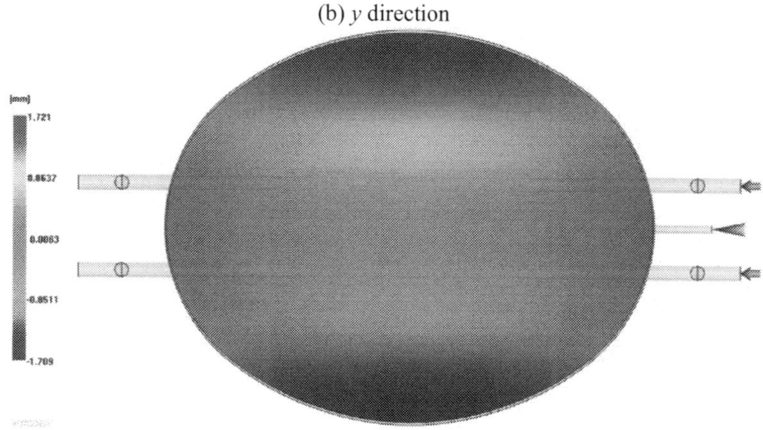

(c) *z* direction

Figure 5.3. *Warpage results. For a color version of this figure, see www.iste.co.uk/zhang/polycarbonate.zip*

Mold temperature (°C)	100
Melt temperature (°C)	300
Injection velocity (cm^3/s)	250
Compression distance (mm)	2
Compression force (t)	1,600
Compression velocity (cm/s)	2
Compression waiting time (s)	15

Table 5.1. *Recommended process parameters for polycarbonate Lexan 141R*

5.1.2. *Residual stress analysis*

During the assembly process, since the warped visor is clamped and forced to deform to its expected shape (as shown in Figure 5.4), residual stresses are generated. In order to correctly describe the residual stresses within the visor, a "reconstruction model" is proposed for the calculation of residual stresses. The procedure is shown in Figure 5.4.

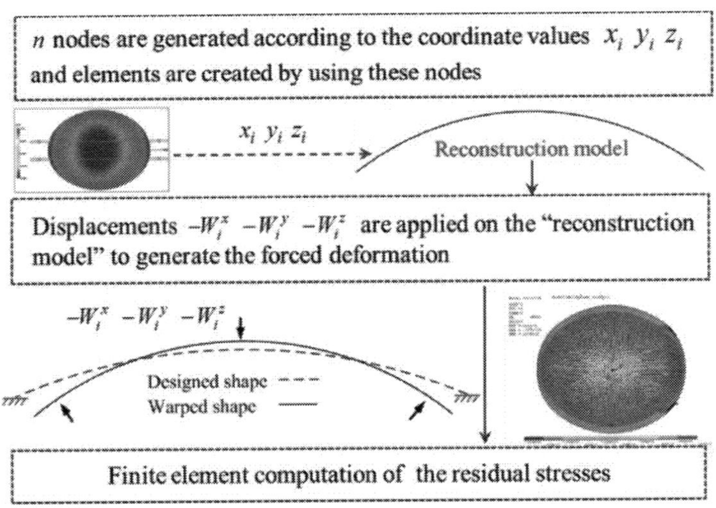

Figure 5.4. *Procedure for computing residual stresses. For a color version of this figure, see www.iste.co.uk/zhang/polycarbonate.zip*

First, n nodes are generated according to the coordinate values x_i, y_i and z_i ($i = 1, 2, \ldots, n$) of the warped visor and elements are created by using these nodes. A new finite element model defined as the "reconstruction model" is obtained based on the generated nodes and elements. Then, a set of displacements with values of $-W_i^x$, $-W_i^y$ and $-W_i^z$ are applied on each node of the "reconstruction model" to generate the forced deformation of the warped visor, and finite element analysis is carried out to compute the residual stresses. The obtained residual stress results will be introduced in the finite element model as the prestress in the following impact analysis.

5.1.3. *Yield stresses and constitutive model*

After the injection molding process simulation, the processing thermal histories of all the elements in the visor are obtained and further exported to predict the yield stresses. Then, the processing-affected yield stress is implemented in the elastic–viscoplastic model of LS-DYNA for further modeling the mechanical properties of the polycarbonate visor.

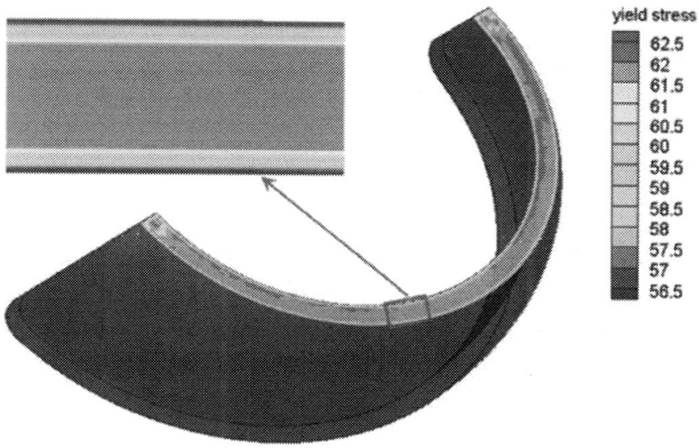

Figure 5.5. *Yield stress distribution in the visor. For a color version of this figure, see www.iste.co.uk/zhang/polycarbonate.zip*

Figure 5.5 shows the distributions of the yield stress in the molded visor. A part of the visor is removed to clearly show the inhomogeneity of the yield stress. A lower yield stress related to the higher cooling rate can be found near the surface when compared to that in the center. In fact, the visor surface contacts with the mold cavity during the processing so that cooling is faster near the surface and slower at the center.

5.1.4. *Impact simulation*

Finite element models consisting of the visor and the impactor are shown in Figure 5.6. The present case corresponds to the spherical debris impact at the central point of the visor. The diameter of the spherical debris is 10 mm. The visor is meshed with 43,200 SOLID 164 elements of LS-DYNA. The spherical projectile is meshed with 28,000 SOLID 164 elements. Note that the meshes of the visor model in the impact simulation are the same as the meshes in the processing simulation.

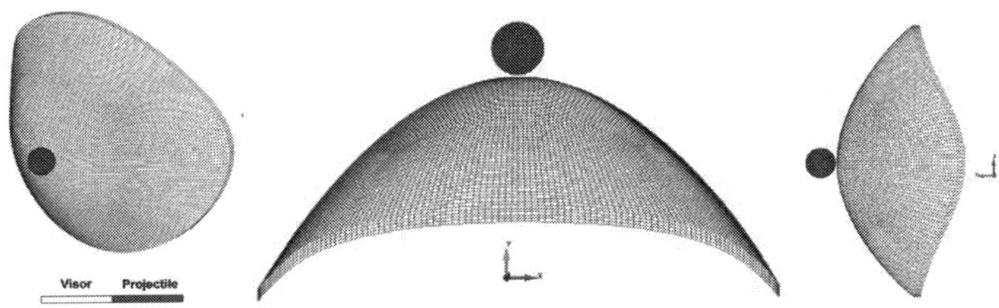

Figure 5.6. *Finite element model of the visor and the impactor. For a color version of this figure, see www.iste.co.uk/zhang/polycarbonate.zip*

The projectile is modeled as a rigid body with a density of 7,800 kg/m^3, Young's modulus of 200 GPa and Poisson's ratio of 0.3. The elastic–viscoplastic constitutive model involving the processing-affected yield stress is used to model the polycarbonate visor. The material properties of polycarbonate, including Young's modulus, Poisson's ratio and density, as well as the parameters of the constitutive model can be found in the previous chapters.

During the impact, the contact between the projectile and the visor is simulated by the contact model **CONTACT_ERODING_NODES_TO_SURFACE*. A strain-based ductile failure criterion is defined through **MAT_ADD_EROSION* to capture the crater and perforation of the polycarbonate visor. The edges of the visor are constrained as a fully clamped boundary condition to avoid any translational and rotational motion. The impact velocity of the projectile is set to 70 m/s.

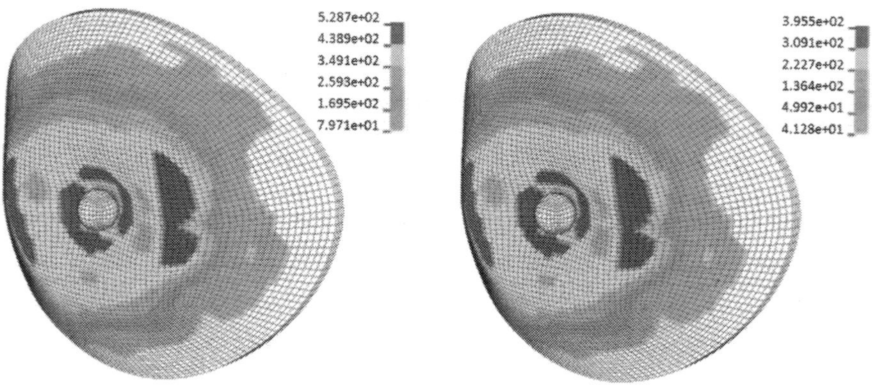

(a) Integrated simulation (b) Simulation neglecting processing influence

Figure 5.7. *Impact simulation results. For a color version of this figure, see www.iste.co.uk/zhang/polycarbonate.zip*

Simulation results of the equivalent stresses of the visor are shown in Figure 5.7. For comparison, the impact simulation neglecting the processing influence is also implemented. The comparison indicates that the injection molding process has a remarkable influence on the mechanical behaviors of the molded product. Although the stress distributions are nearly the same, the difference in the stress magnitude is significant. The maximum stress of the visor computed by the integrated method is 52.87 MPa, while the result neglecting the processing influence is only 39.55 MPa.

5.2. BP neural network model

From the above illustrations, it is found that finite element computations are involved three times in an integrated analysis: injection molding simulation, residual stress analysis and impact simulation. Undoubtedly, costly finite element models are not suitable for a large number of repeated analyses required in an optimization process.

As the artificial neural network has the benefit of high computing efficiency (Fogel and Robinson 2003; Fu *et al.* 2010; Dehghani and Nekahi 2010), a more efficient model created by the BP neural network is proposed to predict the function relationship between the molding process parameters and the maximum von Mises stress of the polycarbonate visor under impact loading. This BP neural network model will be used in the optimization to replace the cumbersome finite element simulations.

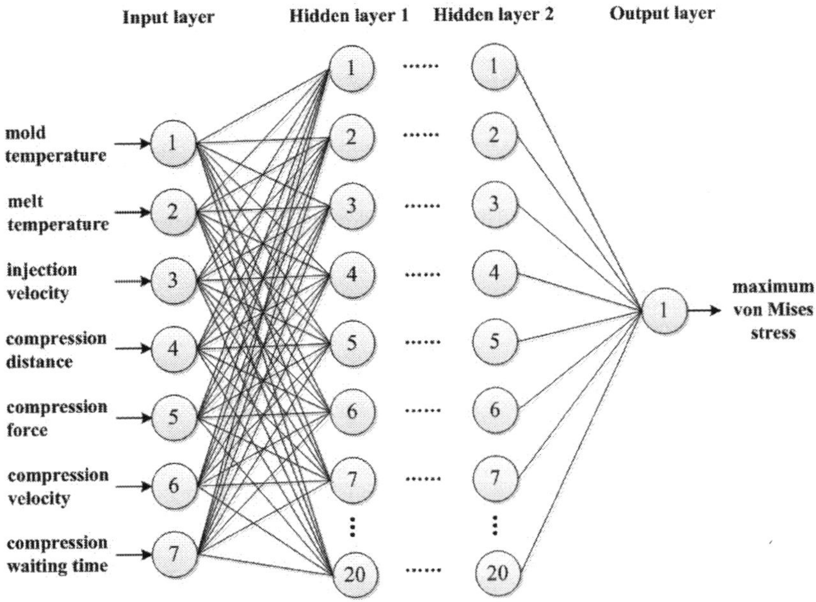

Figure 5.8. *BP neural network architecture used in this study*

It is known that the BP neural network (Fogel and Robinson 2003) has a nonlinear interpolation ability to obtain the mathematical mapping reflecting the inherent mechanism of the experimental data. In this study, a four-layer BP neural network is adopted to generate the function relationship between the process parameters and the maximum von Mises stress that must be minimized in the impacted visor. Every neural network has exactly one input layer and one output layer. We only need to determine the number of hidden layers. Heaton (2008) summarized the capabilities of neural network architectures with various hidden layers: the hidden layer is not needed if the function is linearly separable; one hidden layer can approximate any function that contains a continuous mapping from one finite space to another; two hidden layers can represent an arbitrary decision boundary to arbitrary accuracy with rational activation functions and can approximate any smooth mapping to any accuracy. Therefore, two hidden layers are used in this study. Although two hidden layers increase the computational cost, their ability to represent functions with any kind of shape provides a promising tool for our further study.

The BP neural network architecture used here is shown in Figure 5.8. The network contains three parts: one input layer which has seven neurons related to the mold temperature, the melt temperature, the injection velocity, the compression distance, the compression force, the compression velocity and the compression waiting time; two hidden layers with 20 neurons for each one and one output layer having one neuron representing the mechanical response of impacted visor, i.e. the maximum von Mises stress. There are many rule-of-thumb methods for determining the number of neurons in the hidden layers. However, these rules just provide a starting point for the users to consider. For the problem considered in this study, one of the commonly used rules gives an approximate range for the number of neurons in each hidden layer:

$$n = \sqrt{n_i + n_o} + \alpha \qquad [5.1]$$

where n is the number of neurons in each hidden layer, n_i is the number of neurons in the input layer, n_o is the number of neurons in the output layer and α is a constant from 1 to 10. According to the empirical equation, the number of neurons in the hidden layer is between 4 and 13 in this study. Considering a BP neural network with more neurons in each hidden layer can give higher precision solution (Fogel and Robinson 2003); the number of neurons in each hidden layer is finally selected as 20.

In the network, each neuron receives total input from all of the neurons in the preceding layer as:

$$net_j = \sum_{j=0}^{N} \omega_{ij} x_i \qquad [5.2]$$

where net_j is the total input, N is the number of inputs to the jth neuron in the hidden layer, w_{ij} is the connection weight from the ith neuron in the forward layer to the jth neuron in the hidden layer and x_i is the input from the ith neuron in the preceding layer. A neuron in the network produces its output out_j by processing the input through a transfer function f (the tangent hyperbolic function is used in this study) as follows:

$$out_j = f\left(net_j\right) = \frac{1 - e^{-net_j}}{1 + e^{-net_j}} \qquad [5.3]$$

The BP neural network can be further trained by the data obtained from the integrated simulation before it can accurately predict the maximum von Mises stress of the impacted visor. During the training process, the connection weight w_{ij} is calculated by minimizing the error between the predicted value and the actual value. Details about the training process will be discussed in the following section.

5.2.1. *Generation of training data*

To calculate the connection weights between each neuron, the designed BP neural network should be trained by a group of experimental data. In this study, these data are obtained from the integrated simulations. Each experimental factor range is divided into three levels between the lowest and highest values. The setting of the processing parameters is presented in Table 5.2.

Process parameters	Level 1	Level 2	Level 3
A: mold temperature (°C)	80	100	120
B: melt temperature (°C)	280	300	320
C: injection velocity (cm^3/s)	200	250	300
D: compression distance (mm)	1	2	3
E: compression force (t)	1,400	1,600	1,800
F: compression velocity (cm/s)	1	2	3
G: compression waiting time (s)	14	15	16

Table 5.2. *Setting of the processing parameters*

The experimental data should adequately reflect the inner relationship between the processing parameters and the maximum von Mises stress in the impacted visor. A full factorial experimental design is no doubt an excellent idea. However, the full factorial experimental design means that a large number of computations (3^7 in this study) should be made. Therefore, the Taguchi orthogonal array (Ko *et al.* 1998) using less simulations to find the relationship between parameters is used in this study. Overall, 60 samples, including 18 samples designed by the orthogonal array and other 42 samples randomly generated by computer, are used to train the designed network.

5.2.2. *Training of the neural network*

During the training process, the connection weights should be calculated by minimizing the mean square error between network predictions and training data. According to equation [5.4] (Heaton 2008), connection weights are given by random values at the beginning of the training and then iteratively updated by the gradient method until convergence to certain values:

$$\omega_{ij}^{new} = \omega_{ij}^{old} + \Delta\omega_{ij}$$
$$\Delta\omega_{ij} = -\eta \frac{\partial E}{\partial \omega_{ij}} out_j \qquad [5.4]$$

where E is the mean square error, out_j is the jth neuron output and η is the learning rate parameter controlling the stability and rate of convergence of the network. η is a constant between 0 and 1, which is chosen to be 0.01 in this study.

The mean square error of the training data is set to be 1×10^{-4}. Figure 5.9 shows the training progress of the network. It can be observed that with the updating of the connection weights, the mean square error between the network prediction data and training data decreases gradually and converges to 1×10^{-4} within 4.5×10^5 iterations.

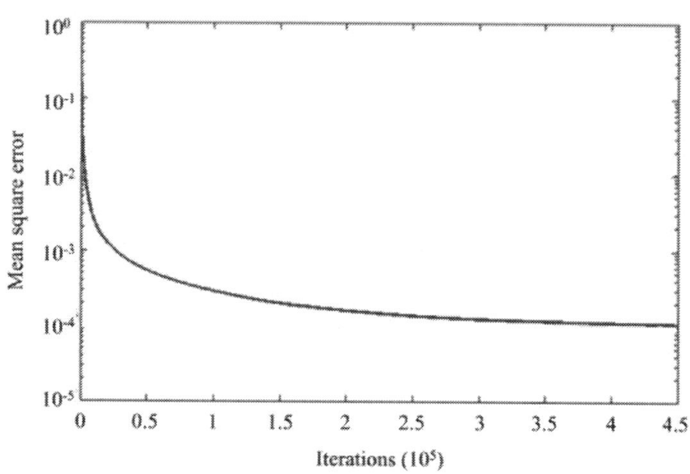

Figure 5.9. *Training process of the network*

After the training, the mathematical mapping between the processing parameters and the maximum von Mises stress is stored in the trained net. The mathematical function can be expressed as:

$$S = f^l\left(\sum w_3 f^s\left(\sum w_2 f^s\left(\sum w_1 X\right)\right)\right) \qquad [5.5]$$

where S is the maximum von Mises stress of the impacted visor; $X = [x_1, x_2, x_3, x_4, x_5, x_6, x_7]$ is the vector consisting of the values of seven different process parameters; f^l is the liner transfer function between hidden layer 2 and the output layer; f^s is the transfer function between the input layer and hidden layer 1, as well as between hidden layers 1 and 2; and w_1, w_2, and w_3 represent the connection weights between the input layer and hidden layer 1, hidden layer 1 and hidden layer 2, and hidden layer 2 and the output layer respectively.

5.2.3. *Neural network testing*

To test the accuracy of the neural network prediction system, 10 groups of processing parameters not used in the training process are employed in the testing. A detailed comparison between the BP neural network prediction model and the integrated finite element simulations is made in Table 5.3.

No.	Processing parameters							Maximum von Mises stress (MPa)	
	A	B	C	D	E	F	G	Integrated simulation	BP neural network model
1	80	280	200	1	1800	3	16	52.5483	52.5880
2	80	300	250	1	1600	1	15	52.5330	53.3044
3	80	300	250	1	1400	2	14	55.2990	54.6517
4	80	320	300	2	1800	3	14	54.0991	53.6957
5	100	280	250	1	1600	3	16	52.1297	52.3264
6	100	300	300	2	1400	2	15	51.0041	51.2704
7	100	320	200	3	1800	1	14	57.9800	56.5787
8	120	280	250	1	1800	1	16	57.9834	57.5685
9	120	300	300	2	1400	2	15	57.4458	56.9829
10	120	320	200	3	1600	14	14	52.5866	51.8862

Table 5.3. *Comparison of BP neural network prediction with integrated simulations*

It can be observed that the BP neural network prediction is in good agreement with the integrated simulation results. This indicates that the developed BP neural network model has a good interpolation ability and can be used as an efficient prediction tool. In particular, the running time of the prediction model is sharply decreased compared to that of the integrated simulation. The average running time of one integrated simulation, including three finite element analyses, is nearly 500 s, while the neural network prediction model only takes about 0.001 s.

5.3. Process optimization by the particle swarm optimization algorithm

5.3.1. *Optimization problem*

The objective of this study is to optimize processing parameters so that the maximum von Mises stress within the impacted polycarbonate visor is reduced. Mathematically, the optimization problem can be formulated as:

$$\begin{aligned}
&Find: A,\ B,\ C,\ D,\ E,\ F,\ G\\
&Minimize: Maximum\ \sigma_{vm}\left(A,\ B,\ C,\ D,\ E,\ F,\ G\right)\\
&80\ ^{\circ}\mathrm{C} \le A: \text{mold temperature} \le 120\ ^{\circ}\mathrm{C}\\
&280\ ^{\circ}\mathrm{C} \le B: \text{melt temperature} \le 320\ ^{\circ}\mathrm{C}\\
&200\ \mathrm{cm^3/s} \le C: \text{injection velocity} \le 300\ \mathrm{cm^3/s}\\
&1\ \mathrm{mm} \le D: \text{compression distance} \le 3\ \mathrm{mm}\\
&1400\ t \le E: \text{compression force} \le 1800\ t\\
&1\ \mathrm{cm/s} \le F: \text{compression velocity} \le 3\ \mathrm{cm/s}\\
&14\ \mathrm{s} \le G: \text{compression waiting time} \le 16\ \mathrm{s}
\end{aligned} \qquad [5.6]$$

The nonlinear implicit nature of the proposed optimization problem induces difficulty in using classical deterministic approaches for solutions. To solve this problem, the particle swarm optimization (PSO) algorithm (Kennedy and Eberhart 1995) is used.

5.3.2. *Basis of the PSO algorithm*

The PSO algorithm is a global optimization algorithm described as sociologically inspired (Kennedy and Eberhart 1995; Fourie and Groenwold 2002). A general illustration of the PSO algorithm is shown in Figure 5.10. In PSO, each individual of the swarm is considered as a particle in a multi-dimensional space that has a position and a velocity. These particles fly through the hyperspace and remember the best position that they have seen. Members of a swarm remember the location where they had their best success and communicate good positions to each other, and then adjust their own position and velocity based on these good positions.

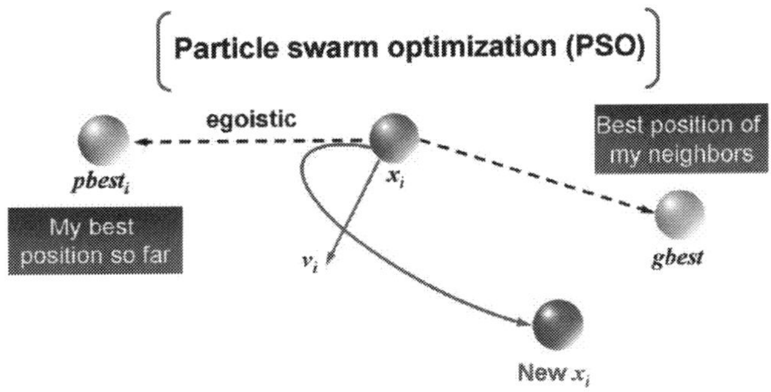

Figure 5.10. *Illustration of the PSO algorithm*

Updating the position and velocity is done as follows:

$$V_i^{k+1} = \omega V_i^k + c_1 r_1 \left(P_i^k - X_i^k\right) + c_2 r_2 \left(P_g^k - X_i^k\right) \quad [5.7]$$

$$X_i^{k+1} = X_i^k + V_i^{k+1} \quad [5.8]$$

where V_i and X_i represent the current velocity and the position of the ith particle respectively; the subscripts k and k+1 refer to the recent and the next iterations respectively; P_i is the best previous position (*pbest*) of the ith particle; P_g is the best global position (*gbest*) among all the particles in the swarm; c_1 and c_2 represent "trust" parameters indicating how much confidence the current particle has in itself and how much confidence it has in the swarm; r_1 and r_2 are two random numbers between 0 and 1; and ω is the inertia weight.

The acceleration constants c_1 and c_2 indicate the stochastic acceleration terms which pull each particle towards the best position attained by the particle or the best position attained by the swarm. Low values of c_1 and c_2 allow the particles to wander far away from the optimum regions before being tugged back, while the high values pull the particles towards the optimum or make the particles to pass through the optimum abruptly. In this work, $c_1 = 2$ and $c_2 = 2$ are chosen.

The role of the inertia weight ω is considered to be important for the convergence behavior of the PSO algorithm. The inertia weight is used to control the impact of the previous history of velocities on the current velocity. Thus, the parameter ω regulates the trade-off between the global (wide ranging) and local (nearby) exploration abilities of the swarm. A proper value of the inertia weight ω provides a balance between the global and local exploration abilities of the swarm, and thus results in better solutions. Numerical tests imply that it is preferable to initially set the inertia to a large value, promote global exploration of the search space and gradually decrease it to obtain refined solutions (Xu *et al.* 2012). Thus, a dynamic variation of inertia weight proposed in (Chang et al. 2010) is used here. The inertia weight ω is decreased dynamically based on a fraction multiplier k_ω as follows:

$$\omega_{k+1} = k_\omega \omega_k \qquad [5.9]$$

5.3.3. *Variable limits handling strategy*

Most optimization problems often include specific constraints and the variable limits. According to equation [5.6], the considered problem belongs to a quasi-unconstrained problem. Only the variable limits, i.e. the design bounds of process parameters, need to be effectively handled to make sure that all of the particles fly inside the search space. A classical way consists of using variable values obtained by random generations satisfying the variable limits. In the present study, an alternative method introduced by Li *et al.* (2007) dealing with the particles that fly outside the variables limits is used.

This method is derived from the harmony search (HS) algorithm. This algorithm is based on natural musical performance processes that occur when a musician searches for a better state of harmony. The engineers seek for a global solution as determined by an objective function, just like the musicians seek to find musically pleasing harmony as determined by an aesthetic (Li *et al.* 2007). The HS algorithm includes a number of optimization operators, such as the harmony memory (HM), the harmony memory size (HMS), the harmony memory considering rate (HMCR) and the pitch adjusting rate (PAR). In the study of Li *et al.* (2007), the harmony memory concept was used in the PSO algorithm to avoid searching trapped in local solutions. The other operators were not used. How the HS algorithm generates a new vector from its harmony memory and how it is used to improve the PSO algorithm is discussed as follows.

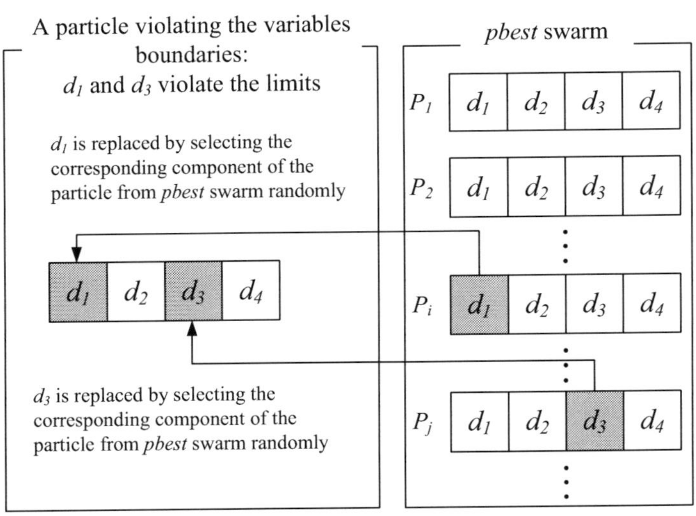

Figure 5.11. *Illustration of the variable limits handling strategy*

In the HS algorithm, the harmony memory (HM) stores the feasible vectors, which are all in the feasible space. The harmony memory size determines how many vectors can be stored. A new vector is generated by randomly selecting the components of different vectors in the harmony memory. Undoubtedly, the new vector does not violate the variables boundaries. When it is generated, the harmony memory will be updated by accepting this new vector if it obtains a better solution and deleting the worst vector. Similarly, the PSO algorithm stores the feasible and "good" vectors (particles) in the *pbest* swarm, as does the harmony memory in the HS algorithm. Hence, the vector (particle) violating the variable limits can be randomly generated again by such a strategy: if any component of the current particle violates its corresponding limits, then it will be replaced by randomly selecting the corresponding component of the particle from *pbest* swarm. Since the new particle is generated based on the feasible and "good" vectors (particles) in the *pbest* swarm, the global solution can be reached in relatively fewer iterations compared with the classical method. To illustrate this strategy, a schematic diagram is shown in Figure 5.11.

5.3.4. *Optimization result*

The optimization problem introduced in section 4.3.1 is solved by using the above PSO algorithm. A population of 50 particles is used. The stopping criterion can be defined based on the number of iterations without updating the best values of the swarm or the maximum number of iterations. Although the latter is not a real physical stopping criterion, it is quite easy in programming implementation and hence widely used in PSO algorithms (Xu *et al.* 2012). In this work, the maximum number of iterations is limited to 1,000 and adopted as the stopping criterion.

The optimization history with iterations for the maximum stress is shown in Figure 5.12. The value of maximum von Mises stress in the polycarbonate visor is finally reduced from 58.2884 to 50.72082 MPa by 12.9% after optimization.

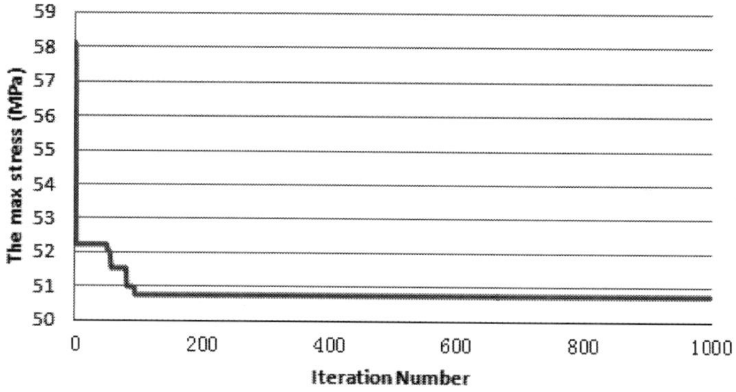

Figure 5.12. *Optimization history with iterations for maximum von Mises stress*

In comparison, the injection molding process parameters recommended by the Moldflow software and the optimized process parameters are listed in Table 5.4. The optimized process parameters have a lower mold temperature, a higher melt temperature, a larger compression distance and compression force as well as a longer compression waiting time. In addition, the compression velocity and the injection velocity slightly increase and decrease respectively.

	Injection molding process parameters						
	A	B	C	D	E	F	G
Recommended parameters	120	280	300	2	1,400	3	15
Optimized parameters	85.366	314.2	293.194	2.7002	1753.6	1.899	15.998

Table 5.4. *Injection molding process parameters before and after optimization*

5.4. Conclusion

In this chapter, an optimization methodology was devised to optimize the injection molding process parameters for high mechanical performance of the product under impact loading. A BP neural network model was constructed to replace expensive finite element analyses in the optimization process. The PSO algorithm was interfaced with the BP neural network model to find the optimal process parameters. The optimization study was finally carried out on a polycarbonate visor under impact loading. The maximum von Mises stress of the initial model was reduced by 12.9%.

In fact, the process optimization of injection molding is mainly limited to the processing parameters to achieve a high dimensional accuracy (Zhou *et al.* 2006; Shen *et al.* 2007; Kurtaran *et al.* 2005; Kurtaran and Erzurumlu 2006; Yin *et al.* 2011a, 2011b; Ozcelik and Erzurumlu 2006) with the warpage reduction of the injection-molded product. A few works have been devoted to the mechanical performance of polycarbonate products. For the design of an injection-molded product, a separation always exists between its process parameters and mechanical performance. Therefore, the proposed optimization methodology provides a promising tool for optimizing the mechanical performance of polycarbonate products from the point view of process parameters.

5.5. References

Chang, N., Wang, W., Yang, W. and Wang, J. (2010). Ply stacking sequence optimization of composite laminate by permutation discrete particle swarm optimization. *Structural and Multidisciplinary Optimization*, 41, 179–187.

Courbebaisse, G. and Garcia, D. (2002). Shape analysis and injection molding optimization. *Computational Materials Science*, 25, 29–35.

Dehghani, K. and Nekahi, A. (2010). Artificial neural network to predict the effect of thermo mechanical treatments on bake hardenability of low carbon steels. *Materials & Design*, 31, 2224–2229.

Fogel, D. and Robinson, C. (2003). *Computational Intelligence: The Experts Speak.* Wiley-IEEE Press, New York, USA.

Fourie, P. and Groenwold, A. (2002). The particle swarm optimization algorithm in size and shape optimization. *Structural and Multidisciplinary Optimization*, 23, 259–267.

Fu, Z.M., Mo, J.H., Chen, L. and Chen, W. (2010). Using genetic algorithm-back propagation neural network prediction and finite-element model simulation to optimize the process of multiple-step incremental air-bending forming of sheet metal. *Materials & Design*, 31, 267–277.

Gu, Y.X., Haimei, L. and Shen, C.Y. (2001). Numerical simulation of thermally induced stress and warpage in injection-molded thermoplastics. *Advances in Polymer Technology*, 20, 14–21.

Heaton, J. (2008). *Introduction to Neural Networks with Java.* Heaton Research, Inc.

Kennedy, J. and Eberhart, R.C. (1995). Particle swarm optimization. *Proceedings of the IEEE International Conference on Neural Networks*, 4, 1942–1948.

Ko, D.C., Kim, D.H., Kim, B.M. and Choi, J.C. (1998). Methodology of preform design considering workability in metal forming by the artificial neural network and Taguchi method. *Journal of Materials Processing Technology*, 80–81, 487–492.

Kurtaran, H., Ozcelik, B. and Erzurumlu, T. (2005). Warpage optimization of a bus ceiling lamp base using neural network model and genetic algorithm. *Journal of Materials Processing Technology*, 169, 314–319.

Kurtaran, H. and Erzurumlu, T. (2006). Effective warpage optimization of thin shell plastic parts using response surface methodology and genetic algorithm. *International Journal of Advanced Manufacturing Technology*, 27, 468–472.

Li, L.J., Huang, Z.B., Liu, F. and Wu, Q.H. (2007). A heuristic particle swarm optimizer for optimization of pin connected structures. *Computers & Structures*, 85, 340–349.

Ozcelik, B. and Erzurumlu, T. (2006). Comparison of the warpage optimization in the plastic injection molding using ANOVA, neural network model and genetic algorithm. *Journal of Materials Processing Technology*, 171, 437–445.

Shen, C.Y., Wang, L.X. and Li, Q. (2007). Optimization of injection molding process parameters using combination of artificial neural network and genetic algorithm method. *Journal of Materials Processing Technology*, 183, 412–418.

Xu, Y.J., Zhang, W.H., Chamoret, D. and Domaszewski, M. (2012). Minimizing thermal residual stresses in C/SiC functionally graded material coating of C/C composites by using particle swarm optimization algorithm. *Computational Materials Science*, 61, 99–105.

Yin, F., Mao, H.J. and Hua, L. (2011a). Back Propagation neural network modeling for warpage prediction and optimization of plastic products during injection molding. *Materials & Design*, 32, 1844–1850.

Yin, F., Mao, H.J. and Hua, L. (2011b). A hybrid of back propagation neural network and genetic algorithm for optimization of injection molding process parameters. *Materials & Design*, 32, 3457–3464.

Zhou, J., Turng, L.S. and Kramschuster, A. (2006). Single and multiobjective optimization for injection molding using numerical simulation with surrogate models and genetic algorithm. *International Polymer Processing*, 5, 509–520.

Index

R, S, T

U, V, W, Y

FSC
www.fsc.org
MIX
Papier | Fördert
gute Waldnutzung
FSC® C083411

Kolibri 360 GmbH
Ferdinand-Jühlke-Straße 7
99095 Erfurt, Deutschland
produktsicherheit@kolibri360.de